Schriftenreihe des Energie-Forschungszentrums Niedersachsen (EFZN)

Band 65

Das EFZN ist ein gemeinsames
wissenschaftliches Zentrum der
Universitäten:

Validation of a two-dimensional model for vanadium redox-flow batteries

Doctoral Thesis
(Cumulative Dissertation)

to be awarded the degree

Doctor of Engineering (Dr.-Ing.)

submitted by

Dipl.-Ing. Maik Becker

from Minden

approved by the

Faculty of Mathematics/Computer Science and Mechanical Engineering,

Clausthal University of Technology

Date of oral examination

21.02.2020

Bibliografische Information der Deutschen Nationalbibliothek

Die Deutsche Nationalbibliothek verzeichnet diese Publikation in der
Deutschen Nationalbibliografie; detaillierte bibliographische Daten sind im Internet
über http://dnb.d-nb.de abrufbar.
1. Aufl. - Göttingen: Cuvillier, 2020
  Zugl.: (TU) Clausthal, Univ., Diss., 2020

D 104

Dean
  Prof. Dr.-Ing. Volker Wesling

Chairperson of the Board of Examiners
  Prof. Dr. rer. nat. Alfred Weber

Chief Reviewer
  Prof. Dr.-Ing. Thomas Turek

Reviewer
  apl. Prof. Dr.-Ing. Ulrich Kunz
  PD Dr.-Ing. habil. Tanja Vidakovic-Koch

© CUVILLIER VERLAG, Göttingen 2020
  Nonnenstieg 8, 37075 Göttingen
  Telefon: 0551-54724-0
  Telefax: 0551-54724-21
  www.cuvillier.de

  ISBN 978-3-7369-7210-0
  eISBN 978-3-7369-6210-1

Abstract

Redox-Flow batteries can play a crucial role in the future electricity supply in order to balance the time lag between the generation of electrical energy from photovoltaics or wind power and the demand for electrical energy. Especially for charge/discharge times under nominal load in the range of 4 to 10 h redox-flow batteries can be a very good storage technology for electrical energy. However, to deploy the technology on a large scale, significant cost reductions are required. A thorough knowledge of the reactions and processes taking place within a redox-flow battery is very helpful for this task and this knowledge can be significantly improved by using a suitable mathematical model to describe the processes in a single cell of a redox-flow battery. Nevertheless, this requires a valid model that has been compared with experimental data and that can reproduce these data plausibly and validly.

In this thesis, the validation of a two-dimensional model for the description of potential and current density distributions in the porous electrodes of a vanadium redox-flow battery is presented, taking into account effects of kinetics and mass transport. Newly developed potential probes are used for *in-situ* measurement of solid and liquid phase potentials within a single cell of a vanadium redox-flow battery. The measurement signals obtained are used for parameter estimation and validation of the model. The probe signals are additionally crosschecked with results from impedance spectroscopy on the same cell and are used to determine the membrane resistance. To improve the meaningfulness of the model, the charge transfer coefficients of the respective reactions between VO^{2+} and VO_2^+ in the positive electrolyte and V^{2+} and V^{3+} in the negative electrolyte are determined in a separate setup by recording current overpotential dependencies. To ensure a homogeneous current density distribution in the electrode of this setup, an adapted, also two-dimensional model is applied to define appropriate experimental conditions.

The results of the measurements indicate the charge transfer coefficients to be in the range between 0.24 and 0.36, thus clearly below a value of 0.5, which is frequently assumed in mathematical models of vanadium redox-flow batteries. The measurements of the membrane resistances of a Nafion® 117 membrane show that the membrane resistance has significantly lower values when charging than when discharging the cell. For example, the area-specific charge resistance in the average state of charge of the electrolyte is only about 0.4 $\Omega \cdot cm^2$, whereas the area-specific resistance during discharge is about 0.7 $\Omega \cdot cm^2$.

In the last section of this thesis, the previous results are integrated in a two-dimensional model in order to describe the experimental measurement data as well as possible with the model. For this purpose, a mass transport parameter is adapted via the modeling environment gPROMS®. The resulting value of the mass transfer parameter is in good agreement with the expected range determined from literature references, and allows this range to be narrowed even further. The description of the measured cell voltage and potential probe signals by the model reveals a good congruence between the model and the experimental data, so the model can be regarded as valid. The model validated in this way can therefore be used very well for the investigation of performance limiting components within the cell, for the optimization of component properties or for the optimization of the operating parameters of a redox-flow battery. Additional suggestions for improvements of the model and the implementation of further models describing membrane crossover effects or electrolyte properties are given.

Kurzfassung

Redox-Flow Batterien können in dem zukünftigen Aufbau der Stromversorgung eines Landes eine entscheidende Rolle spielen, um den zeitlichen Versatz zwischen der Bereitstellung elektrischer Energie aus Photovoltaik oder Windkraft und dem Bedarf an elektrischer Energie auszugleichen. Insbesondere Bei Lade-/Entladezeiten unter nomineller Last im Bereich von 4 − 10 h können Redox-Flow Batterien einen sehr guten Speicher für elektrische Energie darstellen. Um die Technologie jedoch im großen Maßstab einzusetzen, sind deutliche Kostenreduktionen erforderlich. Dafür ist unter anderem ein gutes Verständnis über die ablaufenden Reaktionen und Prozesse innerhalb einer Redox-Flow Batterie sehr hilfreich und dieses Verständnis kann mittels eines geeigneten mathematischen Modells zur Beschreibung der Prozesse in einer einzelnen Zelle einer Redox-Flow Batterie deutlich verbessert werden. Dazu ist allerdings ein valides Modell erforderlich, das mit experimentellen Daten abgeglichen wurde und diese Daten plausibel und valide wiedergeben kann.

In dieser Arbeit wird die Validierung eines zweidimensionalen Modells zur Beschreibung der Potenzial- und Stromdichteverteilungen in den porösen Elektroden einer Vanadium Redox-Flow Batterie unter Berücksichtigung der Kinetik und des Stofftransports aufgezeigt. Dabei werden neuentwickelte Potentialmesssonden zur *in-situ* Erfassung von Fest- und Flüssigphasenpotentialen innerhalb einer einzelnen Zelle einer Vanadium Redox-Flow Batterie eingesetzt. Die dabei erhaltenen Messsignale werden zur Parameteranpassung und Validierung des Modells herangezogen. Die Sondensignale werden zusätzlich mit Ergebnissen aus der Impedanzspektroskopie an derselben Zelle abgeglichen und zur Bestimmung des Membranwiderstands herangezogen. Zur Verbesserung der Aussagekraft des Modells werden die Ladungsübertragungskoeffizienten der betreffenden Reaktionen zwischen VO^{2+} und VO_2^+ im positiven und V^{2+} und V^{3+} im negativen Elektrolyten in einem separaten Aufbau durch die Aufzeichnung von Strom-Überspannungskennlinien ermittelt. Um in diesem Aufbau eine homogene Stromdichteverteilung in der Elektrode zu gewährleisten, wird ein angepasstes, ebenfalls zweidimensionales Modell zur Ermittlung der experimentellen Versuchs-bedingungen eingesetzt.

Die Ergebnisse der Messungen zeigen auf, dass die Ladungsübertragungskoeffizienten im Bereich zwischen 0.24 und 0.36 und somit deutlich unterhalb eines häufig in mathematischen Modellen von Vanadium Redox-Flow Batterien angenommenen Wertes von 0.5 liegen. Aus der Erfassung der Membranwiderstände einer Nafion® 117 Membran ergibt sich, dass der Membranwiderstand beim Laden deutlich niedrigere Werte aufweist als beim Entladen der Zelle. So liegt der flächenspezifische Ladewiderstand im mittleren Ladungszustandsbereich des Elektrolyten bei nur etwa 0.4 $\Omega \cdot cm^2$, dahingegen liegt der flächenspezifische Widerstand beim Entladen bei etwa 0.7 $\Omega \cdot cm^2$.

Im letzten Abschnitt dieser Arbeit werden die vorhergehenden Ergebnisse in einem zweidimensionalen Modell eingesetzt, um die experimentellen Messdaten möglichst gut mit dem Modell beschreiben zu können. Dazu wird ein Stofftransportparameter über die Modellierungsumgebung gPROMS® angepasst. Der resultierende Wert des Stofftransportparameters liegt in guter Übereinstimmung mit dem erwarteten Bereich, der aus Literaturangaben ermittelt wurde, jedoch kann das Ergebnis diesen Bereich noch weiter eingrenzen. Die Beschreibung der gemessenen Zellspannungs- und Potentialmesssondensignale durch das Modell zeigt eine gute Übereinstimmung zwischen dem Modell und den experimentellen Daten auf, sodass das Modell dadurch als valide angesehen werden kann. Das validierte Modell kann daher sehr gut für die Untersuchung leistungslimitierender Komponenten innerhalb der Zelle, für die Optimierung der Komponenteneigenschaften oder zur Optimierung der Betriebsparameter einer Redox-Flow Batterie eingesetzt werden. Weitere Ansätze zur Verbesserung des Modells und zur Implementierung weiterer Teilmodelle, die Membrantransporteffekte oder Elektrolyteigenschaften beschreiben können, werden aufgeführt.

Acknowledgements

I wish to express my sincere appreciation to all those who supported me in my work on this doctoral thesis.

First of all, my deep gratitude goes to my supervisor Prof. Dr.-Ing. Thomas Turek for the opportunity to conduct research in the exciting field of electrochemical energy storage, for the technical and organizational support and the almost endless patience to bring this project to a successful end.

Furthermore, I would like to thank Prof. Dr.-Ing. Ulrich Kunz for the helpful and entertaining discussions, as well as for the manifold ideas on measuring methods and paths for optimization in electrochemistry, and for exciting notes on unexpected reactions within the field of pyrotechnics. I would also like to thank him for the support of the ChemCar competition, as this essentially established my path to the institute and to research.

Special thanks go to PD Dr.-Ing. habil. Tanja Vidakovic-Koch for taking over the third-party review at short notice.

In general, I would like to thank thyssenkrupp Industrial Solutions for financing the project and thus also this work.

I would particularly like to thank thyssenkrupp's project partners, namely Dr. Christoph Roosen and Dr. Bernd Langanke, who initiated this project and brought it to Clausthal University of Technology. My sincere thanks go to Niels Bredemeyer and Nils Tenhumberg for their constructive collaboration and the always very inspiring different view of things. I thank Gregor Polcyn and Peter Toros for the constructive cooperation and the fun hours together.

I thank the former colleagues of the ICVT at the Goslar location for the really great time at the EFZN (now EST), especially during the coffee breaks. My special thanks go to Antonio R. dos Santos for the pleasant and always cheerful atmosphere and the fundamental knowledge in the field of electrochemistry, which he sometimes even shared, Katrin Linke for the processing of numerous organizational "things" and Nadia Kuwertz, who inspired me to use potential probes as well, while she was implementing brittle and tiny dynamic hydrogen reference electrodes in her direct methanol fuel cells. I thank Christine Minke, Jens Riede, Laurens Reining and Katharina Schafner for the relaxed and friendly atmosphere and the nice conversations.

In addition, I would like to thank all colleagues at the ICVT Clausthal location for the many hours spent at and in the institute, both during working hours and at all other times of the day and week, which probably make up an important part of doctoral studies. Above all the duo Rafael Kuwertz and Andreas Köppen, whom I have to express my special thanks to, have had a significant influence on this joyful time.

I would like to thank Jens Friedland for the exciting conversations and the opportunity to generate an exciting scientific idea from the most absurd topics through mutual discussion.

I express my appreciation to Oliver Zielinski and Isabelle Kroner for their pleasant results of their Bachelor theses.

I am very grateful to Karsten and Paul for having persuaded me (and my brother) to study chemical engineering and for having rather advised against a natural science subject.

My big thanks go to my parents for all their support and appreciation, as well as for the freedom and the opportunity since my childhood to pursue technical skills of various kinds.

I thank my twin brother for the discussion of all possible technical and scientific problems, as well as for the possibility to communicate problems and solutions with very few words, as it is probably only very rarely the case.

I owe my sincere thanks to my dearest friend and closest person to me, Marina Bockelmann, who won me over with her charming manner and her cheerful roguishness. Without her support in difficult times, the discussion about the content of my doctoral thesis, the organization and regulation of some apparently everyday activities, the care of our child and the shared political debates, this work would not have been possible.

I would like to thank Tatjana Bockelmann for taking care of my daughter, her granddaughter, at times when I was writing this work.

I would also like to thank all those whom I have forgotten to mention here. Apologies for that.

Contents

1 Introduction

The global energy supply will change in the years to come. "Renewable" energy sources are apparently an essential component to reduce carbon dioxide emissions and to achieve the goals established in the international Paris agreement [1,2]. Fortunately, they have become the most inexpensive energy source for electrical energy in the recent years [3]. Unfortunately, the provision of electrical energy from photovoltaics and wind power is not necessarily linked to the temporal demand for electrical energy. Three options are conceivable and probably a combination of all is essential to overcome this problem. The first option can be assigned to the generic term "smart grid", as one aspect of it deals with the shift in demand for electrical energy at times when it is available. However, this is only a possibility for some electrical consumers and processes, such as heat pumps to provide heat in buildings, charging battery-powered electric vehicles, operating certain domestic equipment (e.g. washing machines, freezer, battery powered devices), but becomes much more complex for industrial processes that rely on continuous production. The second option therefore utilizes the direct storage of electrical energy during an oversupply of solar and wind power in suitable systems that can feed the energy back into the grid at times of need. The conversion of electrical energy into specific compounds (i.e. hydrogen) that can be applied in other sectors, such as industry, mobility, heat or domestic applications are the third variant. Especially the second and third variant possess some interconnections, since for example hydrogen produced by water electrolysis during an oversupply of energy can be used for the production of heat, in fuel cell mobility, as raw material in the chemical industry (third option) or it can be converted back into electrical energy (corresponding to the second option). However, each conversion is correlated with a certain loss of energy, so the path of electricity to hydrogen and hydrogen back to electricity is not as efficient as the storage of electricity in pumped storage hydro power stations or the storage in batteries. Nevertheless, the decision for a single technology or for a whole national or global energy system always implies several issues, such as the energy storage duration (which is the time a fully charged system can release electrical energy at its nominal power), efficiency, power and capacity specific costs, cycle life, expected calendar life and reliability, thus no simple answer can be given.

Some reports have sketched future scenarios that predict the demand of several energy storage and energy conversion technologies for 2050 in Germany. The 85% scenario in a report of the Fraunhofer ISE represents an energy set-up where a reduction of 85% of carbon dioxide emissions related to the year 1990 is achieved [4]. To achieve this goal short duration stationary battery storage systems with a total capacity of 74 GWh, battery storage systems in electric vehicles connected to the grid with a total capacity of 81 GWh and pumped storage hydro power plants with a total capacity of 7 GWh are required for the direct storage of electrical energy from solar and wind power. Additional capacities are required for water electrolysis and methanation to produce energy carrying compounds that allow seasonal storage of energy.

In beginning of 2019 Tennet and Gasunie, a transmission system operator and a company that provides infrastructure and transportation for natural gas, presented an "Infrastructure Outlook 2050" based on a 95% carbon dioxide emission reduction target in 2050 for the Netherlands and Germany [5]. They stated that the share of different storage technologies was very sensitive to political requirements. When considering only the results for Germany, their "local" scenario with its main aim for energy independence, no import from neighboring countries and high amounts of decentral solar power (600 GW) resulted in a significant demand for battery storage (110 GW for Germany) and a high demand for water electrolysis (281 GW). The "national" scenario, aiming for energy independence, with strong supply of offshore wind power and limited energy exchange with neighboring countries corresponds to a comparable demand for water electrolysis (254 GW) but almost no need for battery energy storage (approx. 0 GW).

The most recent report published by Forschungszentrum Jülich comprises two scenarios with a 80% and 95% carbon dioxide emission reduction in 2050 related to the year 1990 [6]. They claimed no need for stationary battery energy storage since already 10% of the capacity of battery electric vehicles applied for grid purposes would be sufficient as short term energy storage. However, for the mid-term storage, their scenarios required compressed air-energy storage capacity of 24 GWh (with 400 MW nominal power, equivalent to 60 h discharge time) for the 80% target and 189 GWh (with 5 GW nominal power, equivalent to 38 h discharge time) for the 95% target. In both scenarios, long term energy storage capabilities are integrated via water electrolysis, methanation and storage of methane and hydrogen in already existing or newly installed methane storage caverns.

These three reports provide the clear message, that seasonal energy storage by water electrolysis with or without succeeding methanation is inevitable for a German economy with a low carbon dioxide footprint. The need for energy storage by stationary batteries, however, is somewhat more uncertain, as some scenarios predict a demand for them while others do not. Nevertheless, any battery technology that could provide short-term (a couple of hours) to mid-term (a few days) energy storage with competitive energy specific costs, efficiency and cycle-life will play an important role in the future energy storage market. The redox-flow battery is generally an energy storage system that can meet these requirements, but a multitude of redox-flow batteries with different electrolyte and electrode compositions have been designed and developed, each with its own characteristics. A brief introduction to redox-flow batteries and its different types can be found in the following section.

1.1 Redox-flow batteries for stationary energy storage

A typical redox-flow battery comprises two vessels that contain dissolved redox-active compounds (mostly salts of metals in different oxidation states), at least one pump for each vessel that pumps the liquid through a stack of single redox-flow battery cells (c.f. Fig 1-1 A). One tank contains a dissolved redox-couple with a low redox potential (N_{red}/N_{ox}), thus defining its electrolyte as the negative electrolyte (NE), whereas the second tank holds the positive electrolyte (PE) with a more positive redox potential of its redox couple (P_{red}/P_{ox}). In each cell two electrodes are flown through by the electrolyte and during charging the oxidized species in the NE are reduced while the reduced species in the PE are oxidized, whereas the reaction is reversed in case of discharging the system. The overall reaction can be described simply as follows:

$$\text{cell reaction:} \qquad N_{red} + P_{ox} \text{ (charged)} \leftrightarrows N_{ox} + P_{red} \text{ (discharged)} \tag{1-1}$$

To prevent mixing of NE and PE both half cells are separated by a separator which can be a porous separator with sufficiently small pores or an ion exchange membrane, that prevents the transport of redox-active species, while still ionically conductive.

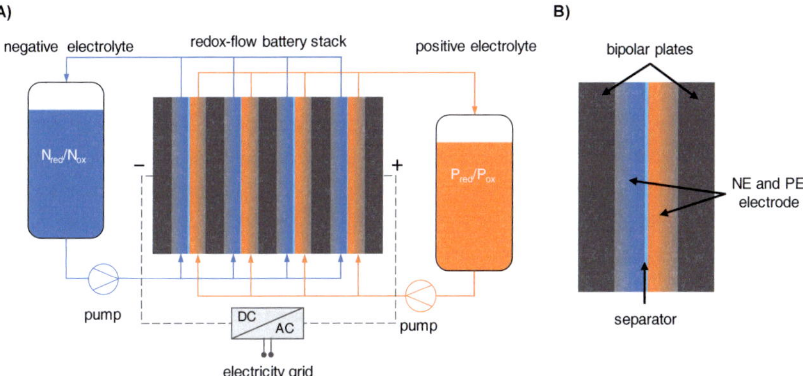

Figure 1-1: A) Schematic illustration of a redox-flow battery with two soluble redox active species in the positive electrolyte (PE) and negative electrolyte (NE), respectively. B) Main components of a single cell, such as bipolar plates separating two adjacent cells, two electrodes where the electrochemical reaction takes place and separator which prevents mixing of both electrolyte but maintains ionic conductivity.

This description of a redox-flow battery fits only to systems where both active species in the NE and PE are dissolved and where no active material is deposited as solid material in the electrodes (e.g. when N_{red} is a metal or when P_{ox} represents an insoluble metal oxide). Therefore, the amount of active species in the system is solely defined by the concentration and the volume of positive and negative electrolyte, while the nominal power is defined exclusively by the characteristics of the redox-flow battery stack (such as cell number, cell area, internal resistance of the cell). In principal, this enables a free scalability of power and energy capacity, which is often mentioned as the advantage of a redox-flow battery. However, even more important is the possibility to achieve low energy capacity specific costs for systems with a large energy capacity to power ratio, since only the costs of active species and additives in the electrolyte as well as costs for electrolyte tanks will dominate the costs of the whole system. Besides these systems with all soluble species, some redox-flow battery types exist, where at least one component is deposited as solid phase in the electrode while charging the system. Further variations exist, that replace one half-cell reaction of an ordinary redox-flow battery by a gas evolving or gas consuming reaction (hydrogen at the negative electrode or oxygen at the positive electrode) during charging and discharging. These systems that combine a solid or gaseous compound in one half cell to an ordinary liquid electrolyte in the other half-cell are classified as hybrid redox-flow batteries [7]. Typical examples of these class with solid compounds in the electrode are the zinc-bromine, the all iron, the all copper or the lead acid battery with soluble lead methanesulfonic acid [8–13]. Despite their use of more or less abundant and inexpensive active materials, the charging reaction in the negative half-cell leads to deposition of a solid metal phase at the electrode. This process always includes the risk of dendrite growth and short circuiting of the cell. In addition, the capacity of the cell is linked to the maximum amount of solid active material that can be stored within the cell, which affects the advantage of individual scalability of power and capacity.

Other hybrid redox-flow batteries make use of hydrogen at the negative electrode and thus they relate quite closely to hydrogen fuel cells. Instead of the slow oxygen reduction or oxygen evolution reaction the positive electrolyte is based on halide/halogen solutions or iron and vanadium salt solutions, that enable typically quite facile reaction kinetics and the storage of at least one charged compound in the liquid phase [7,14,15]. Especially the combination of hydrogen and iron(II)/iron(III) electrolyte could be an interesting option in an energy market that bases already on hydrogen due to its low cost, nontoxic electrolyte and the potential of cost savings, if a future hydrogen grid is used as hydrogen storage unit. Nevertheless, the platinum content that is required at the negative electrode to provide a sufficient catalytic activity could prevent the achievement of low cost targets, as well as the quite low cell voltages might limit its power density and efficiency.

All liquid redox-flow batteries

The class of all liquid redox-flow batteries comprises redox-flow batteries, where both half-cells are flown through by its own liquid electrolyte storing charged and discharged species as dissolved components within. This class contains the majority of different electrolyte chemistries, where several combinations of different positive and negative electrolyte chemistries are possible. Only some of them are depicted and presented here in brief.

Iron-chromium redox-flow battery

The iron-chromium system is one of the earliest redox-flow battery technologies developed at the NASA Lewis Research center and uses a mixed electrolyte that contains dissolved iron and chromium salts in an acidic electrolyte. The electrolyte is applied in both half cells, although two separated electrolyte circuits are used, one for the positive and one for negative electrolyte. The Cr^{2+}/Cr^{3+} redox couple is active in the negative electrolyte while iron just remains as Fe^{2+}. In the positive electrolyte Fe^{2+}/Fe^{3+} is the active redox-couple and chromium remains inactive as Cr^{3+}. The system uses quite inexpensive compounds, and can be operated at high temperatures to approximately 65°C. The overall cell voltage is 1.19 V and the concentration of iron and chromium is approximately 1 M, thus achieving a theoretical energy density of 16 Wh/L. However, the negative half-cell reaction is slow and its redox-potential is quite low, thus hydrogen evolution as a side reaction is likely to occur, reduces the coulombic efficiency and requires a system design that can deal with charge imbalances between positive and negative electrolyte [7,16].

Titanium-manganese redox-flow battery

The titanium-manganese redox-flow battery was developed by Sumitomo Electric Industries and makes use of the relatively abundant elements titanium and manganese [17]. Titanium is varying between its oxidation states Ti^{3+} and TiO^{2+} and manganese between its oxidation states Mn^{2+} and Mn^{3+}, thus leading to a theoretical cell voltage of 1.41 V. However, the charged Mn^{3+} species tend to undergo disproportionation reaction and precipitate as MnO_2, which can lead to loss of capacity and potentially more severe damages by blocking the electrodes of the redox-flow battery. To mitigate this issue, the concentration of sulfuric acid and the concentration of titanium salts must be sufficiently high, while the concentration of manganese should not exceed 1 M. Such a setup yielded an energy density of 24 Wh L^{-1}, but the high redox-potential of the manganese redox couple leads to carbon corrosion. Since the MnO_2 precipitation still occurs at elevated states of charge leading to mass transport limitations and because the kinetics at least for the negative half-cell are too slow, further development is necessary to bring this type of redox-flow battery onto the market [18–20].

Quinone based organic redox-flow batteries

A first metal free redox-flow battery was presented by Huskinson et al., who applied 9,10-anthraquinone-2,7-disulfonic acid as the redox active component in the negative electrolyte and bromine in the positive electrolyte in an acidic aqueous electrolyte [21]. The quinone (and quinones in general) undergo a very rapid electron transfer reaction, enabling quite high power densities and its easy production from commodity chemicals anthraquinone and oleum offers the potential for low costs. However, the open circuit voltage in range of 0.7 to 0.9 V is quite low, and bromine/bromide as the positive redox couple comprises certain risks due to its toxicity, environmental hazards and high vapor pressure. To overcome this risk, Yang et al. dissolved 1,2 benzoquinone-3,5-disulfonic acid in diluted sulfuric acid to be used as the positive electrolyte [22]. As negative active component anthraquinone-2-sulfonic acid was applied, which is a similar approach to Huskinson et al. The experimental open circuit cell voltage decreased to values of roughly 0.5 V and the solubility of the organic components was quite low, hence they mentioned to introduce suitable functional groups into the quinone molecules to adapt solubility and cell potential. Accordingly, Lin et al. introduced more replaced sulfonic acid functional groups by hydroxy functional groups in the anthraquinone molecule, thus lowering its redox potential [23]. To maintain solubility, they used an alkaline electrolyte of 1 M potassium hydroxide solution, that even further decreased the pH dependent redox-potential of the anthraquinone redox couple. As positive electrolyte, they applied ferro/ferricyanide dissolved in 1 M potassium hydroxide. The cell voltage increased to 1.2 V and even higher voltages of up to 1.34 V were indicated by using anthraquinone with more hydroxy or methyl functional groups. Eventually, quinone based redox-flow batteries seem to offer some promising potentials for energy storage, but solubility and long term stability issues need to be addressed.

All-soluble all iron redox-flow battery

A redox-flow battery system that bases solely on the redox reaction of iron was published by Gong et al. who added triethanolamine to the negative electrolyte, which serves as organic-ligand to iron, lowering the redox potential of the Fe^{2+}/Fe^{3+} couple significantly. They used the ferro/ferricyanide couple in the positive electrolyte, while both electrolytes consisted basically of diluted sodium hydroxide solution (3 M). Their setup demonstrated a good performance at open circuit cell voltages of about 1.3 V and a quite stable discharge capacity over 110 cycles. Unfortunately, crossover of triethanolamine from the negative to the positive electrolyte across the membrane was identified as on major challenge as it increases the membrane resistance and can affect the properties of the positive electrolyte. Nevertheless, their work highlighted the possibility to adapt redox-potentials of redox couples by appropriate metal-organics ligands, which is not limited to iron.

All vanadium redox-flow battery

The most mature and most often used redox-flow battery was developed by Maria Skyllas-Kazacos in 1986 [24–26]. The aqueous electrolyte comprised of dissolved vanadium salts (concentration of 0.1 M) and 2 M sulfuric acid. The negative electrolyte contains $V^{2+}/\ V^{3+}$ species, while the positive electrolyte contains VO^{2+}/VO_2^+ species during charge and discharge. Although the first cell setup-was quite simple it still achieved coulombic efficiencies of more than 90% up to a state of charge of 70%, corresponding to a quite stable open circuit cell voltage of 1.3 V. In the following years, their group could show several improvements such as cell design, electrode pretreatment, electrolyte production and electrolyte stability [27–32]. One basic advantage of vanadium is its property to exist in four oxidation states, whereas two couples maintain a sufficient redox-potential difference and are soluble in aqueous electrolyte. Therefore, a vanadium redox-flow battery (VRFB), is unharmed by any cross contamination of vanadium from one side to the other, since this simply leads to a self-discharge reaction, but does not affect the system irreversibly. This property is quite exceptional among the elements of the periodic table and only a few distinctively more hazardous elements have comparable properties (Cr^{2+}/Cr^{3+} vs. Cr^{3+}/CrO_4^2, Np^{3+}/Np^{4+} vs. NpO_2^+/NpO_2^{2+})[7].

This list gives just a brief overview of some important redox-flow battery types. However, more combinations of different active species and electrolytes are already reported and the reader is referred to the literature for a more detailed overview. Just to name a few, other systems are semi-solid flow batteries [33], polymer-based organic redox-flow batteries [34], differential pH redox-flow batteries [35] and different aqueous and non-aqueous organic redox-flow batteries [36,37].

1.2 Vanadium redox-flow battery – basic properties

The vanadium redox-flow battery (VRFB) is the most mature redox-flow battery system and comprises several advantages compared to other redox-flow-batteries. In this thesis the VRFB is used as model system to develop new characterization and model validation methods for redox-flow batteries. Therefore, a brief description of its main components is presented.

Vanadium electrolyte

As already mentioned the electrolyte contains the four different vanadium species, dissolved in a sulfuric acid electrolyte. The typical vanadium concentration ranges from 1.5 M to 2 M vanadium whereas the total vanadium concentration lies about 4 M [32]. The temperature range is depending on state of charge as well as on vanadium and acid concentration. The upper temperature limit is restricted by the thermal stability of VO_2^+, since it tends to undergo an irreversible precipitation reaction building solid V_2O_5 particles. To prevent this precipitation the maximum temperature limit is often set to a value of about 40 °C, while higher concentrations of acid and lower SoC can increase this limit [38]. The lower temperature limit is attributed to the limited solubility of V^{2+} and V^{3+}, that start to crystallize at temperature lower than 15 °C, for a total sulfate concentration of 3 M and a vanadium concentration of 1.6 M [39]. Lower concentration of sulfuric acid increase the solubility and allow the decrease of the lower temperature limit. A mixed acid electrolyte using sulfuric and hydrochloric acid could widen the usable temperature window and increased the solubility of vanadium, but overcharging of the positive electrolyte would lead to chlorine gas evolution increasing the need for safety features to prevent overcharging and to handle potential evolution of chlorine gas [40].

The standard redox potentials of the main reactions in the positive and negative electrolyte (NE, PE) can be found in [41]. The main reactions in the NE and PE, as well as the overall cell reaction are as follows (discharge direction from left to right):

negative electrolyte: $\qquad V^{2+} \leftrightarrows V^{3+-} + e^-$ $\qquad E_{ref,NE} = -0.255$ V vs. SHE $\qquad$ (1-2)

positive electrolyte: $\qquad VO^{2+} + H_2O \leftrightarrows VO_2^+ + 2\,H^+ + e^-$ $\qquad E_{ref,PE} = 1.004$ V vs. SHE $\qquad$ (1-3)

overall cell reaction: $\qquad V^{2+} + VO_2^+ + 2\,H^+ \leftrightarrows V^{3+} + VO^{2+} + H_2O$ $\qquad E_{OCV} = 1.26$ V vs. SHE $\qquad$ (1-4)

The theoretical overall open circuit cell voltage E_{OCV} is 1.26 V, however practical cell voltages indicated higher values, that are probably related to activity effects and some sort of Donnan potential across the membrane [42]. The electrolytes conductivity at room temperature is higher for the PE ranging from about 290 to 410 mS cm^{-1} and the NE range is between 180 and 250 mS cm^{-1}, whereas both are depending on state of charge and temperature [43]. The electrolyte's viscosity lies in a range of about 4 to 6 mPas for room temperature condition and 1.6 M vanadium and about 4 M total sulfate concentration [44].

The main cost driver within the vanadium electrolyte is vanadium itself, since demineralized water is practically unlimited available and sulfuric acid is a commodity chemical. Vanadium is traded as ferrovanadium and as vanadium pentoxide, with only the pentoxide is used for electrolyte production due to its lower costs and its high vanadium amount without iron contamination. The price per kg vanadium pentoxide from 2000 to November 2019 is given in Fig. 1-2 and additionally the corresponding price per kWh of energy storage capacity is given. Since not the whole amount of vanadium can be utilized within a vanadium redox-flow battery, we assumed a vanadium utilization of 70% and a discharge voltage of 1.3 V (realistic only for high performance systems or low discharge current densities). The average value is around 12 € kg^{-1} vanadium pentoxide which equals capacity specific costs of about 91 € kWh^{-1}. The U.S. Department of Energy set a cost target of 150 $ kWh^{-1} for energy storage technologies [45], hence for vanadium redox-flow batteries about two third of the cost target are already consumed by the high costs of the vanadium pentoxide. This indicates a certain need to optimize the whole system in terms of power density and resulting cost-reduction, as well as a need for quite high energy capacity to power ratios, since this lowers the effect of additional cost of cell stacks, pumps and the whole balance of plant.

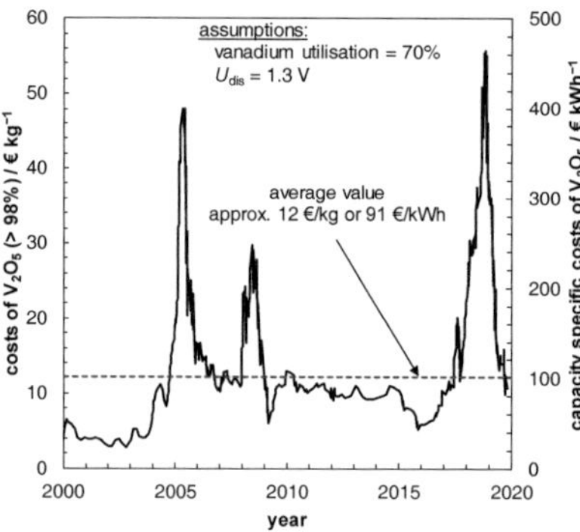

Fig. 1-2: Historical development of the costs for vanadium pentoxide and its corresponding costs for a kWh of energy storage capacity in a vanadium redox-flow battery system (cost of vanadium pentoxide based on [46] and exchange rate for USD to EUR taken from [47]).

The global mine-production of vanadium in 2018 was about 73·10^3 metric tons [48]. The biggest market share in 2018 had China with a value of 55%, followed by Russia (25%), South Africa (12%) and Brazil (9%). The global annual production corresponds to a theoretical maximum energy capacity of 24 GWh, which is just a third of the total required battery energy capacity for Germany in 2050 in scenario presented by Henning & Palzer [4]. When considering the global reserves and resources of vanadium, that are estimated to 20 million and more than 63 million metric tons [48] theoretical capacities of 6.6 TWh and 21 TWh are achievable. Although this would be theoretically more than sufficient for Germany's requirement of energy storage capacities, it is clear that this is an unrealistic approach and vanadium redox-flow batteries can be just one part amongst other energy storage systems in a future energy supply.

VRFB system components

A typical vanadium redox-flow battery system consists of two tanks, one for the NE and another one for the PE. Two pumps feed the electrolyte via appropriate pipes and hoses through the cell stack, where the charging and discharging reactions take place and the electrolyte then returns to the tank (c.f. Fig 1-3). The tanks are sealed and often purged with inert gases such as nitrogen or argon since the V^{2+} species undergo oxidation in contact with oxygen from air (or other sources), which needs to be prevented. Obviously, all components in contact to the electrolyte require corrosion resistant materials, due to the acidity of both electrolytes and their quite high and low redox-potentials.

The core component of the system is the cell, where the main reactions take place and which determines the main losses that are inevitable during charging and discharging. Losses in other system parts can be kept quite low, such as stray-current losses in the cell electrolyte inlet/outlet hoses that lead to self-discharge, losses by providing pumping power, conversion losses in the AC-DC converter or losses in auxiliary equipment for battery management, safety-features and temperature control. The cell itself comprises basically four components: a bipolar plate that ensures separation between two adjacent cells, two electrodes typically porous carbon felt or carbon fiber electrodes and a separator.

The bipolar plate is typically the mechanical backbone of a single cell, that guarantees a sufficient mechanical stability. Additionally, it provides electrical contact between two adjacent cells, while preventing electrolyte contamination between these two cells. This can be achieved with compound materials based on mainly graphite and a low amount of polymer, such as polypropylene. For large cells or for cells that require low pressure drop the bipolar plate can include a flow field, which is a combination of several channels integrated into the bipolar plate that distribute the electrolyte across the cell.

The porous electrodes, which are often carried out as carbon felt electrodes must maintain a high surface area to reduce reaction overpotentials, while still ensuring a good electrical connection to the bipolar plate. Furthermore, the microscopic surface should provide a good wettability with the electrolyte and should facilitate the reaction by appropriate functional groups or defect sites on its surface. Nevertheless, the electrolyte must be pumped through the electrode, so a high permeability is necessary to limit the pressure drop across the cell inlet and outlet to an acceptable value.

The separator provides ionic conductivity between both electrodes, while still separating both electrolytes, preventing cross-mixing and self-discharge of the cell. It can be executed as an anion exchange membrane, a cation exchange membrane or a microporous separator. Ion exchange membranes are usually hydraulically tight and allow only the selective migration of cations (mainly protons) or anions (sulfate anions) across the cation or anion exchange membrane, respectively. They cannot prevent the permeation of V^{2+} or V^{3+} from the NE side to the PE side or the permeation of VO^{2+} or VO_2^+ from the PE side to the NE side completely, hence this crossover leads to a certain self-discharge of the cell, that needs to be minimized. A microporous separator provides no specific selectivity and is permeable for the electrolyte. However, the permeability should be sufficiently low, so that any pressure fluctuations between negative and positive half-cell lead only to a minor electrolyte crossover. Small pores in the nanometer scale might have a beneficial effect on vanadium-crossover, since they could reduce the diffusivity of vanadium species within the pore, if they are just small enough.

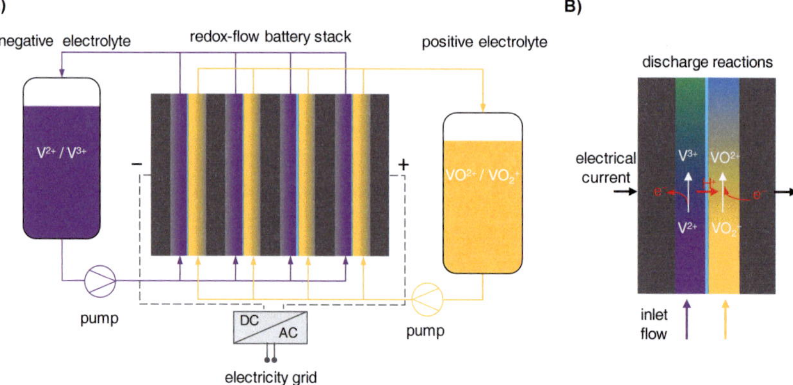

Figure 2-3: A) Schematic illustration of a vanadium redox-flow battery system and its components. B) Rough illustration of the reactions and electronic and ionic currents within a single cell of a vanadium redox-flow battery.

As already mentioned a low-cost design is mandatory to achieve the given cost target of the U.S. Department of Energy (150 $ kWh^{-1} [45]). Since the costs for the vanadium electrolyte are presumably in a range of about 100 $ kWh^{-1}, it is essential for the redox-flow battery stack to deliver as much power with as less investment as possible. This in turn, can be achieved whether by building a low-cost stack with moderate power densities, or by building a stack for moderate costs but achieving high power densities. Therefore, besides knowledge of the cost footprint of several components, an in-depth knowledge of the behavior of each component and its impact on power density, when modifying its key parameters is crucial. For this task modelling of redox-flow batteries is a helpful tool to understand the behavior of the redox-flow battery and to identify the limiting components that need improvement.

A couple of models has already been presented, however model validation is and crucial task to acquire trustworthy results from the model. A brief overview of models and model validation is given in the following section.

1.3 Overview of mathematical modelling and model validation

Modeling literature

Several works were published dealing with different cell models and especially different dimensionalities describing the electrode processes [49]. Zero-dimensional (0D) models, one-dimensional (1D), two-dimensional (2D), and three-dimensional (3D) models were applied by several groups and have been applied for different investigations. Zero-dimensional models are often applied in investigations where low computational effort is required, e.g. for the online estimation of SoC and remaining capacity in an operating RFB [50–54] or for the calculation of stray currents [55–59]. Another field for the application of 0D models is the modeling of slow mass transport effects due to membrane crossover [60–64], of total system behavior and system efficiency optimization with a special focus on flow rate optimization [65–73] or of thermal modeling of whole RFB systems and investigation of temperature fluctuations [74–78]. The general applicability of 0D models for RFBs was verified by Sharma et al. who identified two constraints for the dimensionless conversion rate and the interchangeability number [79]. These constraints are mostly fulfilled for small lab cells, however for larger lab cells of 100 cm² and commercial cells with even larger cell areas it is unlikely that they fulfill the constraints for interchangeability number.

Instead of 0D models 1D models are less frequently applied, but Vynnycky et al. and Chen et al. demonstrated the accuracy of a 1D model with dimensions in the through plane direction (perpendicular to electrolyte flow) [80,81] and considered its limits at high current densities for low flow rates or low SoC. Lei et al. applied a 1D model for the investigation of membrane transport properties and potential distribution within the membrane including Donnan potential [82] and an ion selective adsorption process [83]. Sun et al. showed the possibility to fit a 1D impedance model to experimental data to determine losses caused by different mass transport and charge transfer processes within the cell [84]. Based on this work Pezeshki et al. and Zago et al. extended the investigation method [85,86], Pezeshki et al. by investigating two different electrode materials with two different flow designs, Zago et al. by adopting the model to a more numerical solution estimating the steady state, linearization at the steady state and finding the impedance spectra by Laplace transformation. 1D plug flow models (model dimension in direction of flow) were published by Li et al. 2016 who investigated effects of cell voltage monitoring placement, electrolyte imbalance and flow rate optimization [87]. Koenig et al. used a comparable model, but focused on the current density distribution along the flow direction and the resulting overpotentials, so he could render the results of Li et al. more precisely [88].

The number of publications dealing with 2D RFB is large compared with papers considering 1D models and this could be attributed to the good compromise of computation time and accuracy. 2D models comprise the dimensions that are used in the 1D plug flow and 1D through plane models and do not contain the third dimension due to symmetry conditions. Though, they are able to evaluate concentration and current density distributions in both dimensions, and e.g. allow identification of areas in the cell, where the main charge transfer reaction is taking place. Walsh's group started recent publication history with a couple of models dealing with a steady state isothermal model, that was later adapted with the ability to describe non-isothermal operation, side reactions, and dynamic processes [41,89–92]. You et al. included a flow rate dependent mass transfer coefficient in their model [93] that bases on mass transport equations determined on carbon fiber electrodes by Schmal et al. [94] and this was applied in the majority of published models later on. Knehr & Agar et al. implemented effects of membrane crossover due to convection, migration and diffusion and were therefore able to validate and predict capacity fade [95,96]. Bromberger et al. published a model, that calculated the conductivity of the electrolyte based on an empirical equation [43], instead of calculating it by a modified Nernst-Planck equation [97]. They investigated the effect of electrode compression and made a sensitivity analysis on the kinetic parameters describing the electrode reactions. However, no model validation was given in their work and the description of mass transport from the bulk phase to the carbon fiber electrode was simplified. A comparable approach was published by Zhou et al., who calculated accurate electrolyte conductivities by application of Stokes-Einstein equation with concentration dependent ion-mobilities [98]. This results in a more appropriate prediction of main reaction zones and concentration distributions. Rudolph et al. applied a constant parasitic discharge of V^{2+} ions to calculate for charge imbalance instead of implementing vanadium crossover and showed a good agreement between modeling and experimental energy and charge demand of the RFB [99]. Xu added the effect of different viscosities on the cell performance, since the viscosities for the positive and negative electrolyte differ and depend on the state of charge of the electrolytes and thereby influence mass transport effects [100]. An anion exchange membrane was described by the model Wandschneider et al. published in 2014 [101], leading to transport of bisulfate ions instead of protons across the membrane. Khazaeli investigated the optimum flow rate for yielding a high total system efficiency [102]. Yang et al. investigated the effect of membrane crossover by diffusion, migration and electro-osmotic convection [103]. They could estimate the concentration and potential distribution across the electrolyte, the membrane and their interfaces, while describing Donnan potential jumps and total vanadium crossover fluxes. An increase of electro-osmotic transport should mitigate the effect of capacity fade of a vanadium RFB. A similar investigation was done by Darling et al., who analyzed migration and electro-osmotic transport in comparison to diffusion for several membranes [104]. Their key findings were the proportional dependence of migration and electro-osmotic transport on current density compared to diffusional transport that is inversely proportional to membrane thickness. The effects of flow variation and different total vanadium concentrations on discharge behavior and overpotentials were investigated by another group that applied a 2D RFB model [105]. In a following paper Yang et al. identified an optimum of round trip efficiency for medium current densities, since high current densities lead to high losses in the cell and low current densities lead to increased pumping losses of RFB system [106]. Furthermore, a SoC-dependent current density can extend available capacity and discharge energy, while maintaining an almost constant round trip efficiency.

A number of 3D models are reported, that are dealing with comparable effects and investigations as 2D models do. However, 3D models are advantageous for pore scale modeling [107,108], since this would be hardly feasible in 2D models. CFD modelling of flow field designs coupled with electrochemical reactions is another field in which 3D models can help to understand the performance limitations of certain cell design [109–112]. Other aspects, such as dynamic behavior, non-isothermal modeling and description of membrane crossover processes [113–118] do not necessarily benefit from 3D models.

Validation in literature

Besides the potentials and progresses in RFB modeling, validation is indispensable to confirm congruence between experimental and calculated results. Though it is rather unclear, why sometimes validation has not been carried out at all or only very rudimental. An often-applied validation method is the modelling of a single charge discharge cycle under constant current conditions and adaption of the model parameters to fit the cell voltage to the experimentally recorded cell voltage. Sometimes the operation parameters are extended (flow rate and/or current density) to improve the experimental basis for validation. More intensive validations were carried out by only a few groups.

As already mentioned Sun et al. and Zago et al. applied models describing the impedance spectra of a RFB, giving a good opportunity to validate the model in terms of kinetic, ohmic and mass transport losses [84,86]. Wandschneider et al. validated their 2D model to the cell voltage during nine consecutive charge discharge cycles with current densities between 0.25 and 0.75 $kA \cdot m^{-2}$ [101]. Although their work shows an adequate agreement between model and experimental data for low current densities certain deviations appear for current densities over 0.6 $kA \cdot m^{-2}$. König et al. validated the modeled stack voltage to the experimental data of a commercial system under 6 different operating conditions by fitting the scaling factor in their model [119]. Since the cell setup was not disclosed the results are of limited use to optimize the cell performance. Pugach compared his modeled data to experimental results of three discharge polarization curves at a constant state of charge but different flow rates and further on validated his model to match four charge discharge curves at current densities between 0.4 to 1.0 $kA \cdot m^{-2}$ [120]. Gandomi applied four liquid phase potential probes in the positive electrode of a vanadium RFB [121], as already done by their group before [122,123], and the resulting information of liquid phase potentials during discharge was used to validate his model, showing a good agreement. However, Gandomi used the model to predict and discuss concentration distributions within the electrode and it remains questionable if potential probe sensing is an appropriate tool to validate these. A publication of Mazur et al. showed the potential of pseudo-reference electrodes in determining the effects of carbon felt pretreatment in-situ on both half-cell reactions [124]. Although Mazur did not used it for model validation, the authors of this work think, that it is a good method to validate a model. Vanadium redox-flow batteries can be an efficient component for the energy storage in a prospective energy market. However, better understanding is required for further optimization and modeling of redox-flow batteries can be an efficient tool to achieve this. Modeling of redox-flow batteries requires an adequate model validation to have a realistic approximation of the behavior of a redox-flow battery and to identify the bottle-necks for its performance.

In this thesis, the application of new developed potential probes, the determination of charge transfer coefficients in an *ex-situ* setup under controlled current density conditions and the application of these results for parameter estimation and model validation of a two-dimensional redox-flow battery model is demonstrated. The thesis is separated in four main chapters, whereas the first three are presented as published publications and the fourth represents a not yet submitted manuscript. All four chapters describe specific parts of the whole work:

- **Chapter 3** introduces a new kind of potential probes based on laminated carbon fibers. These solid phase potential measuring probes are applied during polarization curve measurements in a vanadium redox-flow battery and the evaluation of dynamic and quasi-steady state allows determination of specific electrical resistances as well as current density distribution for three different carbon felt compression rates.

- **Chapter 4** describes an *ex-situ* setup capable of acquiring overpotential current density dependence for thin carbon felt and carbon paper electrodes in vanadium electrolyte under operating conditions that where designed in terms of superior mass transport and sufficiently even transfer current density distribution. The resulting information allowed the extraction of charge transfer coefficients subsequently used in the following chapters.

- **Chapter 5** focusses on the evaluation of overpotential data gathered by solid and liquid phase potential probes applied in a vanadium redox-flow battery during polarization curve and impedance spectroscopic measurements. The results enabled a crosscheck of overpotential and impedance spectroscopic measurements, as well as the presentation of kinetic reaction rate constants, area specific resistances of the membrane and bipolar plate resistances including contact resistance effects.

- Eventually, in **Chapter 6** the results of the preceding chapters are combined and the experimental results are integrated into a modelling environment that enables parameter estimation of mass transport coefficients and validation of a mathematical model describing the electrode processes within a redox-flow battery.

2 Overview of publications

The following publication were submitted and published in "peer-reviewed" scientific journals and are an integral part of this doctoral thesis.

1. M. Becker, N. Bredemeyer, N. Tenhumberg, T. Turek, Polarization curve measurements combined with potential probe sensing for determining current density distribution in vanadium redox-flow batteries, J. Power Sources, 307 (2016) 826-833.

   The 5-year impact factor of the Journal of Power Sources is **6.823 (2018)**. The contribution of the first author was the preparation and execution of the experiments, as well as data evaluation and the writing of the manuscript. This publication is reprinted in Chapter 3 of the thesis and describes the general setup of the cell, the test stand and the solid-phase potential probes to run polarization curves of vanadium redox-flow batteries. The solid-phase potential probes are applied to maintain information on the homogeneity of the solid-phase potential and dynamic and quasi steady-state signals of the probes can be evaluated to evaluate information on the resistivity of the carbon felt carbon bipolar plate sandwich, as well as information on the current density distribution within the cell.

2. M. Becker, N. Bredemeyer, N. Tenhumberg, T. Turek, Kinetic studies at carbon felt electrode for vanadium redox-flow batteries under controlled transfer current density conditions, Electrochim. Acta, 252 (2017) 12-24.

   The 5-year impact factor of the Journal Electrochimica Acta is **4.940 (2018)**. The contribution of the first author was to design and to operate an experimental setup, as well as evaluation of received data and mathematical modelling of the porous electrodes. This publication is reprinted in chapter 4 of the thesis and describes a parameter variation with the help of the mathematical model, that guarantees appropriate experimental conditions to achieve reliable data for the charge transfer coefficients of both reactions within a vanadium redox-flow battery. These charge transfer coefficients were applied in the evaluation of further measurements and allowed a more precise setup of following models of the vanadium redox-flow battery.

3. M. Becker, T. Turek, Combination of impedance spectroscopy and potential probe sensing to characterize vanadium redox-flow batteries, J. Power Sources, 446 (2020) 227349.

The 5-year impact factor of the Journal of Power Sources is **6.823 (2018)**. The contribution of the first author was the preparation and execution of the experiments, as well as data evaluation and the writing of the manuscript. This publication is reprinted in Chapter 5 of the thesis and demonstrates an experimental approach to measure overpotentials within a redox-flow battery with the help of liquid and solid phase potential probes inserted in the cell. An additional evaluation of impedance spectroscopic data allowed the determination of reaction rate constants and exchange current densities, that could be checked by the overpotentials measured by the probes. Furthermore, membrane and bipolar plate resistances were experimentally accessible within an operating redox flow battery, giving the opportunity to use this data for more precise model validation and investigation on mass transport.

3 Polarization curve measurements combined with potential probe sensing for determining current density distribution in vanadium redox-flow batteries

Journal of Power Sources 307 (2016) 826–833

Contents lists available at ScienceDirect

Journal of Power Sources

journal homepage: www.elsevier.com/locate/jpowsour

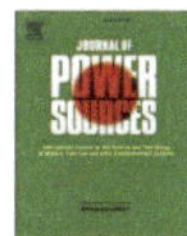

Polarization curve measurements combined with potential probe sensing for determining current density distribution in vanadium redox-flow batteries

Maik Becker [a, *], Niels Bredemeyer [b], Nils Tenhumberg [b], Thomas Turek [a]

[a] Clausthal University of Technology, Institute of Chemical and Electrochemical Process Engineering, Leibnizstr. 17, Clausthal-Zellerfeld 38678, Germany
[b] ThyssenKrupp Industrial Solutions AG, Friedrich-Uhde-Str. 15, Dortmund 44141, Germany

HIGHLIGHTS

- Solid phase potential probes are applied to a vanadium redox-flow battery (VRFB).
- A method for determining the felt resistance of the VRFB is presented.
- Effect of different carbon felt compression rates on felt resistance is shown.
- Known felt resistances allow the determination of localized current densities.

GRAPHICAL ABSTRACT

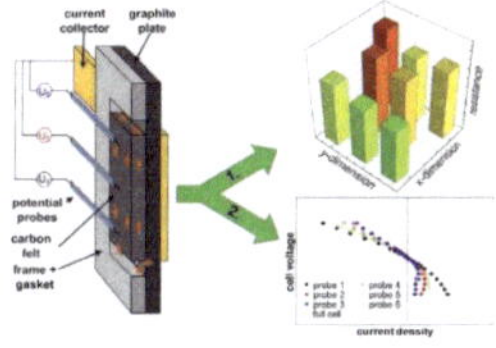

ARTICLE INFO

Article history:
Received 22 September 2015
Received in revised form
23 November 2015
Accepted 1 January 2016
Available online 1 February 2016

Keywords:
Vanadium redox-flow battery
Potential probes
Carbon felt resistance
Current density distribution
Carbon felt compression
Electrolyte flow rate

ABSTRACT

Potential probes are applied to vanadium redox-flow batteries for determination of effective felt resistance and current density distribution. During the measurement of polarization curves in 100 cm^2 cells with different carbon felt compression rates, alternating potential steps at cell voltages between 0.6 V and 2.0 V are applied. Polarization curves are recorded at different flow rates and states of charge of the battery. Increasing compression rates lead to lower effective felt resistances and a more uniform resistance distribution. Low flow rates at high or low state of charge result in non-linear current density distribution with high gradients, while high flow rates give rise to a nearly linear behavior.

© 2016 Elsevier B.V. All rights reserved.

1. Introduction

The change in the production of electrical energy from fossil fuels to renewable energy sources increases the demand for electrical energy storage systems compensating the intermittent nature of renewable energy sources. Redox-flow batteries among others are promising devices due to their capability of separately customizing power and energy capacity by cell design and tank size, respectively. Although a number of redox couples are possible the all vanadium redox-flow battery (VRFB) is the most common one [1–3]. The VRFB electrolyte contains vanadium in four different oxidation states used as electrochemically active species that

* Corresponding author.
E-mail address: maik.becker@tu-clausthal.de (M. Becker).

http://dx.doi.org/10.1016/j.jpowsour.2016.01.011

3 Polarization curve measurements combined with potential probe sensing for determining current density distribution in vanadium redox-flow batteries

undergo redox-reactions during charging or discharging. While charging V^{3+} is reduced to V^{2+} at the negative electrode and VO^{2+} is oxidized to VO_2^- at the positive electrode. The reverse reactions take place during discharging:

$$V^{2+} \rightleftharpoons V^{3+} + e^- \tag{1}$$

$$VO^{2+} + H_2O \rightleftharpoons VO_2^+ + 2H^+ + e^- \tag{2}$$

Although commercial VRFBs are already available, further cell improvement is necessary to reduce the power specific costs of the system. Besides new developments in the field of electrocatalysts [4–6], membranes [7–12] and electrolytes [13–16], flow field design and flow strategies become more important. Ma et al. studied the effect of switching the flow rate from a low to a high value, if the voltage increases to a specific threshold during charging or discharging, on the system efficiency [17]. Tang et al. modeled concentration overpotentials and pressure drop losses within a flow battery stack of 40 cells and found that a variable flow rate with a flow factor of 7.5 yielded the highest system efficiencies [18]. Rudolph et al. calculated the flow distribution within a multi inlet cell. The highest utilization was achieved using only one inlet and two outlet channels at the opposite endings of the cell [19].

Different flow field designs were investigated in several studies [20–25]. Xu et al. compared cells with serpentine and parallel flow fields to cells without flow fields by three dimensional modeling. The serpentine flow field design resulted in an even overpotential distribution across the cell, a good mass transport into the felt and a low pressure drop across the cell [20]. Yin et al. developed a three dimensional model to describe an interdigitated redox-flow battery, that predicts a homogenous flow distribution if the land width and flow rate become sufficiently high.

As modeling needs experimental validation, the measurement of current density distribution is essential to confirm homogenous current density and overpotential. While this is a relatively common tool in fuel cell research [26,27], there are only a few publications dealing with current density distribution in redox flow batteries.

Hsieh et al. used shunt resistors with gold-plated current collector segments or divided segments within the bipolar plate to measure the current density distribution during charging and discharging experiments [28]. The measurement setup using divided segments of the bipolar plate seemed to be beneficial for a higher accuracy by decreasing the amount of lateral currents but also lowered the cell performance, due to higher contact resistances. Clement et al. used a printed circuit board with segmented shunt resistors for determining the current density in each segment [29]. These authors used a serpentine flow field bipolar plate which was partially segmented by machining a grid from the back side of the bipolar plate. This reduced the lateral currents and allowed a precise measurement of current density distributions within cells made of different carbon paper electrodes. In a split cell setup the difference between a heat treated electrode and an untreated electrode containing 5% of PTFE was very significant. However, the expensive setup, the susceptibility to corrosion in case of electrolyte leakages and the challenging application to large cells are drawbacks of these methods.

Our approach is the development of a more robust, inexpensive method to measure the current density distribution based on potential probes. Aaron et al. used dynamic hydrogen reference electrodes to investigate the in-situ kinetics of a VRFB [30]. A quite similar setup was applied by Ventosa et al. through sandwiching a silver–silver sulfate reference electrode between two layers of membrane separating the positive and negative electrode [31]. Liu et al. studied the electrolyte phase potential distribution in the positive electrolyte using platinum wires covered by a layer of PTFE [32], which allowed to localize the main reaction zone within the porous electrode. The potential probes used in these three works allowed to measure the liquid phase potential and to determine overpotentials as well as ionic resistances. These experimental data are viable for improving the kinetics but they are not necessarily helpful to check for a homogeneous current density distribution, since the liquid potential is a function of felt resistance, electrolyte resistance, mass transport, kinetics and, depending on the setup, of membrane resistance.

Therefore we decided to measure the solid phase potential by probes based on long carbon fibers that were inserted into the cell as an array. The solid phase potential only depends on the local felt and monopolar plate resistance as well as on the local current density distribution. The dependency can be described using Ohm's law (3), where the potential difference φ is a result of the local current density j and the sum of the electrical resistances of felt R_{felt} and monopolar plate $R_{monopolar\ plate}$, respectively.

$$\varphi = \left(R_{felt} + R_{monopolar\ plate} \right) \cdot j \tag{3}$$

Because the measured solid phase potential is neither directly influenced by electrolyte resistance, mass transport, kinetics nor membrane resistance, it provides more accurate and reliable data than measurement of the liquid phase potential. Furthermore, the method is very inexpensive and easily applicable to large cells or cell stacks.

2. Experimental

2.1. Material

The vanadium electrolyte (1.6 mol/L vanadium concentration and 4 mol/L total sulfate concentration) was purchased from GfE (Gesellschaft für Elektrometallurgie, Germany). The initial electrolyte was a 1:1 mixture of V(III) and V(IV) resulting in a state of charge (SoC) of −50%. The sulfuric acid 95%–98% (Ph. Eur.) and phosphoric acid 85% (Ph. Eur.) purchased from Carl-Roth (Germany) was diluted with Milli-Q water (18.2 MΩ cm) to concentrations of 1 M. Volumetric standard solutions of 0.02 M potassium permanganate were purchased from Carl Roth. 0.1 M Fe(II) solutions were prepared using ammonium iron (II) sulfate hexahydrate (p.a., Carl Roth) dissolved in 1 M sulfuric acid.

2.2. Cell construction

The cells consisted of 7 mm thick PPG86 monopolar plates (Eisenhuth GmbH & Co. KG, Germany) and a single layer of untreated GFD4.6 EA carbon felt (SGL Carbon GmbH, Germany) with uncompressed dimensions of $100 \times 100 \times 4.6$ mm^3 resulting in an active cell area of 100 cm^2. Silicon gaskets and poly-methylmethacrylate (PMMA) frames of different thicknesses were used to assemble three cells with carbon felt compression rates of 9%, 28% and 42%, respectively. The gaskets and frames had inner dimensions of 120×100 mm^2 to provide a sufficient supply channel of 10 mm in height for an even electrolyte distribution across the cell. The inlets were placed on the bottom left and the outlets were placed at the top right position of the cell. Two 2 mm thick brass plates served as current collectors connected to the monopolar plates via a gold plated nickel mesh to reduce contact resistances. The positive and negative electrodes were separated by a Nafion® 117 membrane. Prior to cell assembly the membrane was swelled for at least 24 h in 1 wt.-% sulfuric acid. Two 22 mm thick end plates of PMMA yielded a homogeneous compression of the carbon felt and permitted a back fed electrolyte supply as well as

M. Becker et al. / Journal of Power Sources 307 (2016) 826–833

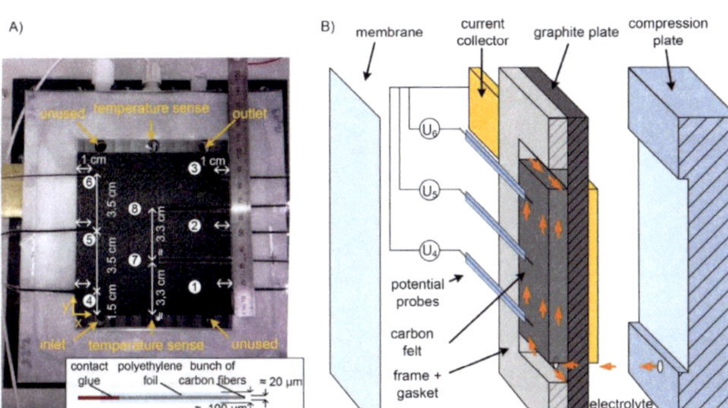

Fig. 1. A) Picture of the positive half cell with potential probes lying on the carbon felt. The inset shows the schematic setup of a single potential probe. B) Cross section view on the positive half cell.

the insertion of four Pt-100 resistance temperature sensors (Class A, Omega Engineering, Inc., USA). Eight solid phase potential probes were distributed across the positive electrode having contact to the carbon felt and to the membrane as well (Fig. 1). Probes 1 to 3 were placed 1 cm from the right edge of the felt and probes 4 to 6 symmetrically distributed 1 cm from the left edge of the felt. The distance between the electrodes on each side was 3.5 cm. Probes 7 and 8 were arranged horizontally in the center of the felt with a space of 3.3 cm between them. The potential probes were made of a bunch of carbon fibers (Toolcraft, Germany) that was sealed within two layers of lamination foil (made of PE), each 40 µm thick. The resulting probe was cut in 5 mm wide and approximately 20 cm long stripes having a bunch of uncovered fibers at both ends. At one end of the stripe the empty space between particular fibers was filled with contact glue (Turbocoll 2000, Turbo Kleber Plastik, Boldt & Co. OHG, Germany) to prevent a leakage of electrolyte along the fibers. At the other end the uncovered fibers were cut to a length of approximately 5 mm to have a sufficient contact area to the carbon felt. The end tightened with contact glue was later used outside the cell by connecting to the auxiliary voltage inputs of the test stand, while the other end was contacted to the carbon felt inside the cell. To ensure proper positions of the potential probes and to facilitate the cell assembly, the potential probes were fixed on the silicon gasket by using small stripes of adhesive tape (tesafilm®, Germany). Since the carbon fiber material of the potential probe is very similar to the carbon felt fibers and the number of fibers of each potential probe is very small, the effect of the potential probe fibers on the current distribution is negligible. The voltage drop across the felt was measured between each potential probe and the current collector of the positive electrode, while the cell voltage was measured between both current collectors. The typical thickness of the whole potential probe and the uncovered bunch of carbon fibers at the end was approximately 100 µm and 20 µm, respectively.

2.3. Electrochemical characterization

The characterization of three different cells ran on a fully automated test stand delivered by FuelCon AG, Germany. Two gear pumps propelled the electrolyte through the cell, while the

pressure drop and flow rate across each half cell were recorded. The conductivity and temperature of the electrolyte were measured in the conduits of both sides. The electrolyte vessels, each containing approximately 4 L, were equipped with a heating jacket that was flown through by water tempered at 25 °C. Auxiliary inputs recorded the temperature of the four temperature probes and the eight potential probes within the cell. A pure nitrogen stream for each half side removed evolving gases and prevented oxidation of V(II).

At the beginning of each experiment the electrolyte was balanced by remixing positive and negative electrolyte in order to achieve a SoC of nearly −50%. Subsequently the electrolyte was charged applying a high flow rate of 333 mL/min at a constant voltage of 1.7 V until a SoC of almost 95% was reached. The SoC of both electrolytes was confirmed by redox titration. After charging the cell, the characterization process started using electrochemical impedance spectroscopy at flow rates from 36 mL/min up to 491 mL/min. The impedance spectroscopic data was described with an equivalent electrical circuit model presented by Mohamed et al. [33] which considers the internal ohmic cell resistance in series with two RC networks representing the electrode reactions. In this work only the ohmic cell resistance was of interest. After recording impedance data at several flow rates, the flow rate was reduced to 36 mL/min and the first polarization curve was recorded by applying different potentials to the cell. Each potential step endured for 60 s achieving a current density close to steady state but still maintaining only a small change of the electrolytes SoC. After measuring the open circuit potential (OCP) of the cell for 60 s followed by application of an initial potential of 1.3 V, the potential was alternated around 1.3 V with potential steps of ±50 mV in the beginning and ±700 mV in the end leading to a minimum and maximum voltage of 0.6 V and 2.0 V, respectively. Alternating between low and high potentials yields charging and discharging currents every two steps and is therefore beneficial compared to linear polarization, since the change of the SoC is quite small [30]. Every second all signal inputs were recorded. Subsequent to finishing the first polarization curve at a high SoC, the flow rate was changed to 56 mL/min, 95 mL/min, 175 mL/min, 333 mL/min and 491 mL/min and a complete polarization curve was recorded at

M. Becker et al. / Journal of Power Sources 307 (2016) 826–833

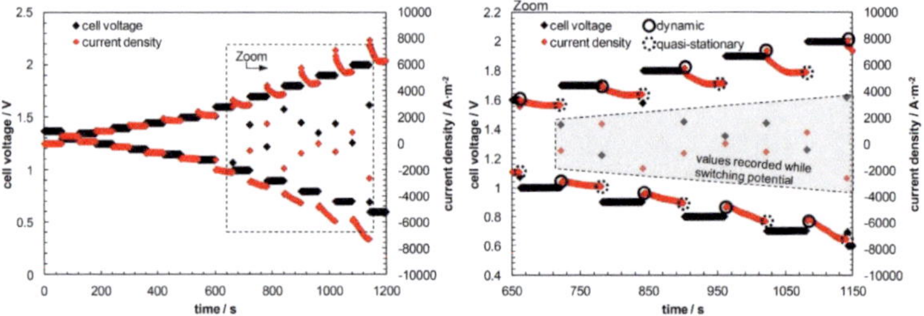

Fig. 2. Applied alternating cell voltage and resulting current density at a flow rate of 491 mL/min and a state of charge of 35% (carbon felt compression rate 42%). The zoomed area is marked in the left plot. In the zoom on the right solid circles exemplarily mark the dynamic current density directly after switching the potential. Dotted circles exemplarily mark the quasi-steady state current density after keeping the potential constant for 60 s. Single potentials and current densities plotted in the grayed box were recorded during switching the potential between two potential steps. They were omitted from evaluation.

each specific flow rate. This impedance spectroscopic and polarization curve sequence was repeated four times starting at different SoC of roughly 75%, 50%, 25% and 10%. Between these sequences the electrolyte was discharged at 0.8 V and a flow rate of 491 mL/min. After reading the data for the last polarization curve an electrolyte sample was taken and the very low SoC of positive and negative electrolyte were determined by redox titration. To calculate the correct SoC of positive and negative electrolyte during the measurements the conductivity and temperature signals within the supply conduit of the cell were applied to the empirical equation proposed by Corcuera et al. [34].

$$SoC = \frac{\kappa - C \cdot T - D}{A \cdot T + B} \qquad (4)$$

Here, κ is the conductivity of the electrolyte in mS/cm, T the temperature of the electrolyte in °C and A, B, C and D are empirical constants. D and B were adapted to fit the high and low SoC values determined by redox titration for each electrolyte before and after the characterization. The SoC of the complete system was calculated as the average of the SoC of positive and negative electrolyte.

2.4. Electrolyte titration

The SoC was determined using a redox titration method based on the work of Treadwell et al. [35]. A precise volume of negative electrolyte was diluted in 1 M ortho-phosphoric acid to a total vanadium concentration of ca. 25 mM and titrated with potassium permanganate solution (0.02 M) at a temperature of at least 80 °C. The redox potential was recorded between a platinum ring electrode and a silver/silver chloride electrode (Metrohm AG, Switzerland). The whole titration run automatically on a Titrando 888 automated titrator system (Metrohm AG) under pure nitrogen atmosphere to prevent oxidation of V(II). During titration three potential jumps appear according to the different oxidations states of vanadium, that allow to calculate the SoC and the total vanadium concentration. The positive electrolyte was diluted in 1 M sulfuric acid to a concentration of ca. 25 mM and subsequently titrated at room temperature against the same reference electrode. Prior to titration with potassium permanganate a sufficient excess of Fe(II) solution (0.1 M Fe(II) as ammonium iron(II) sulfate in 1 M sulfuric acid) was added to reduce all V(V) species to V(IV). The first potential jump allows the calculation of remaining Fe(II) and therefore the SoC, while the second one determines the amount of the total vanadium in the electrolyte.

3. Results and discussion

The time dependent signals of cell voltage and current density obtained from the polarization curve measurement are shown in

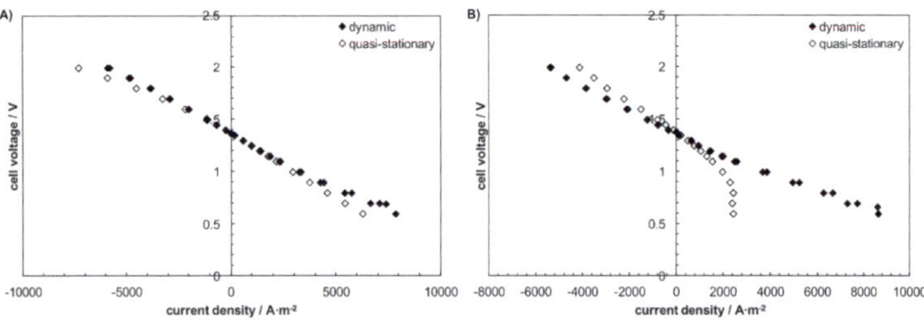

Fig. 3. Polarization curves under dynamic and quasi-stationary conditions (carbon felt compression rate 42%). A) flow rate of 491 mL/min and SoC of 35%, B) flow rate of 37 mL/min and SoC of 39%.

M. Becker et al. / Journal of Power Sources 307 (2016) 826–833

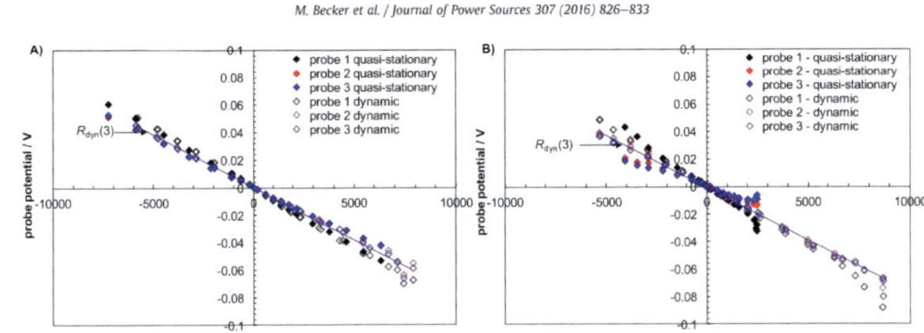

Fig. 4. Current density and corresponding probe potentials (shown only for probe 1 to 3) according to polarization curves shown in Fig. 3 A) flow rate of 491 mL/min and SoC of 35%, B) flow rate of 37 mL/min and SoC of 39%. Linear regression is used to determine the dynamic carbon felt resistances as it is exemplarily shown for probe 3.

Fig. 2. The controlled voltage signal alternates around 1.3 V leading to intermittent charging and discharging currents. Directly after switching from charging to discharging, the current density rises to a maximum followed by a smooth decrease and a quite stable quasi-stationary signal after 40–60 s. This is the expected behavior since the cell was charged prior to discharging, which results in an increase of charged species on the surface of the carbon fibers. At the beginning of the discharging process these surface species promote the reaction significantly but due to their limited amount their concentration rapidly declines and the ongoing discharge leads to an increasing concentration gradient within the cell that reduces the current density. We assume that this decline is not caused by a change of the concentration at cell inlet, because of the high electrolyte volume in the electrolyte vessel. After switching from discharge to charge, an unexpected continuous decrease of the negative charging current occurs (meaning an increase of the absolute value of the charging current), which is until now not fully understood. Neither temperature effects nor inner mass transport effects from the bulk of the electrolyte to the carbon fiber surface seem to be responsible for this behavior, since the temperature decreases by less than 3 K during charging under all circumstances while different flow rates show the unexpected continuous decrease of current density during charging in different intensities. At high flow rates of 491 mL/min the continuous decline is very noticeable at current densities from -2000 A/m^2 to -8000 A/m^2 (s. Fig. 2), while lower flow rates of 175 mL/min to 56 mL/min show a lesser decrease of current density. At the lowest flow rate of only 36 mL/min the current density rises during charging and shows no decline. This is the expected behavior that can be explained by the ongoing decrease of discharged species during charging. However, due to higher flow rates the inner mass transport should be enhanced, but this would cause only a faster increase to a constant current density level but not a constant decrease as observed.

Therefore other effects such as ion-crossover across the membrane due to migration and electro-osmosis as described by Darling et al. and Yang et al. [36,37], that are especially important at high current densities, might be responsible for the reported behavior.

Nevertheless the current density after a period of 60 s was assumed to be in a steady-state condition and is hereinafter referred to as "quasi stationary" current density. The current density recorded directly after switching and reaching the new potential will be referred to as "dynamic" current density. At high flow rates the difference between dynamic and quasi stationary data is marginal (Fig. 3), since the concentration gradient within the cell is reduced by the intense convection for the quasi stationary condition. Low flow rates increase the difference due to electrolyte utilization at high current densities, but both dynamic data show a good agreement. For this reason it can be assumed that the determined dynamic data are free of mass transport limitations, because switching of current ensures a sufficient concentration of the reacting species on the surface of the carbon fibers, independent of the position within the carbon felt.

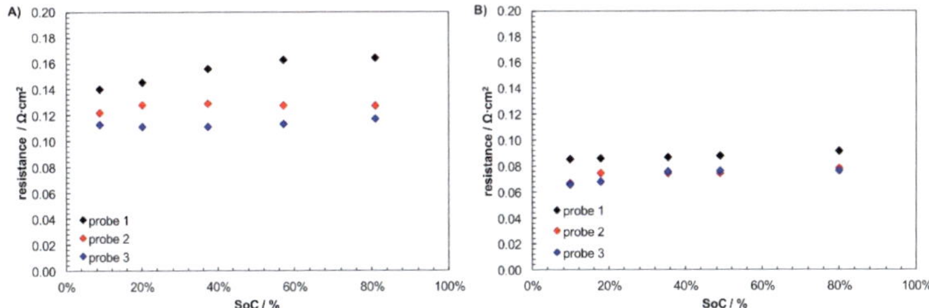

Fig. 5. Felt resistances of probe 1 to 3 at different SoC and a flow rate of 491 mL/min. A) For a carbon felt compression rate of 9% and B) for a carbon felt compression rate of 42%.

23

M. Becker et al. / Journal of Power Sources 307 (2016) 826–833

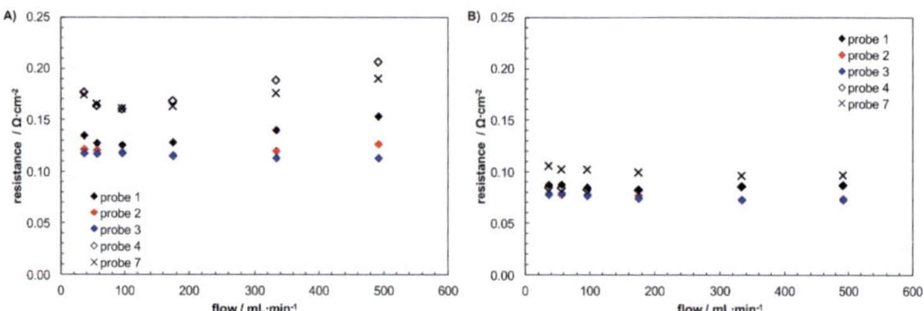

Fig. 6. Felt resistances of probe 1 to 3 at different flow rates averaged over SoC. A) For a carbon felt compression rate of 9% and B) for a carbon felt compression rate of 42%.

The separation in dynamic and quasi-stationary data done for the potential probe signals results in data presented in Fig. 4 (exemplary for probes 1, 2 and 3 distributed over the height of the cell close to the right). The dynamic data shows a linear dependence of probe potential on current density, meaning that its behavior is dominated by electrical resistance according to Ohm's law (s. equation (3)). Furthermore the slopes of the dynamic signals of probe 1, 2 and 3 are very similar. The quasi stationary signals are non-linear at low flow rates and the potential drop of a potential probe positioned close to the outlet (i.e. probe 3) declines while the potential drop of a probe close to the inlet increases at high current densities, which is caused by mass transport limitations and electrolyte utilization. Since the recorded dynamic data is supposedly independent of mass transport limitations and because felt resistance is negligible in comparison to the total cell resistance, an even current density distribution in the entire cell can be assumed for dynamic conditions. Thus the slope of the probe potential versus the dynamic current density allows the determination of a dynamic

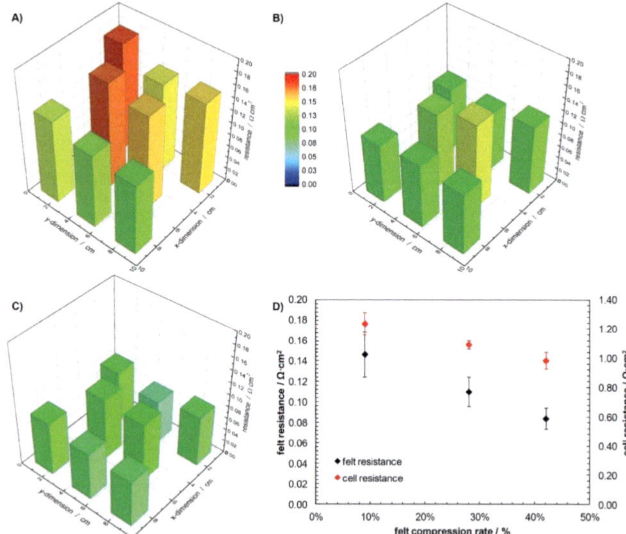

Fig. 7. Averaged localized felt resistances of three cells with carbon felt compression rates of A) 9%, B) 28% and C) 42%. D) Averaged felt resistances as a function of felt compression rate compared to the total cell resistance determined via impedance spectroscopy.

M. Becker et al. / Journal of Power Sources 307 (2016) 826–833

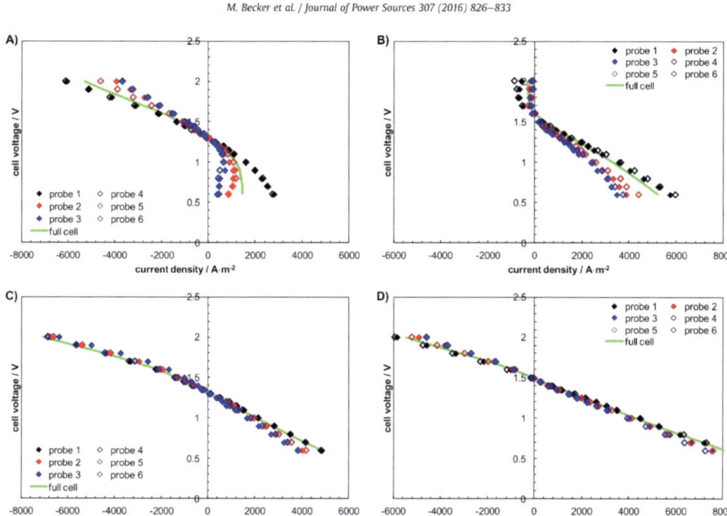

Fig. 8. Polarization curves with localized current densities for a carbon felt compression of 42%. A) Flow rate of 56 mL/min and a SoC of 14%. B) Flow rate of 56 mL/min and a SoC of 96%. C) Flow rate of 491 mL/min and a SoC of 18%. D) Flow rate of 491 mL/min and a SoC of 80%.

resistance R_{dyn}. It comprises the resistances of the current collector, the monopolar plate and the carbon felt electrode. As the specific electrical resistances of the current collector (<0.1 μΩm, estimated) and the monopolar plate (550 μΩm, provided by manufacturer) are considerably lower than that of the carbon felt (<5 mΩm, provided by manufacturer), the measured resistance is mainly affected by the felt. However, it should be noted that the measured effective felt resistance is not directly comparable to *ex situ* measured felt resistances, since the felt electrode is porous and the current density within the felt continuously decreases down to zero on its path from the graphite plate to the membrane. Therefore the measured potential drop and the resulting effective felt resistance are lower, compared to values obtained by *ex situ* measurements.

The dynamic resistances obtained from all polarization curves were calculated using linear regression and least square fit. They are mainly independent of SoC (Fig. 5) and show a decrease for higher compression rates. Potential probes 1, 4 and 7 show a dependence on flow rate for low compression rates (Fig. 6 A). These probes exhibit a minimum for a flow rate of 95 mL/min which might be attributed to the position of these probes close to the electrolyte inlet. High flow rates and turbulent flow regimes possibly due to an inhomogeneous flow distribution in this area could affect the fiber–fiber interconnections within the carbon felt and might lead to a higher effective felt resistance. At higher carbon felt compression rates the number of fiber–fiber interconnections increases, hence a dependence on flow rate is no longer observed (Fig. 6 B). The averaging of dynamic resistances for all flow rates and all SoC leads to localized felt resistances shown in Fig. 7. With increasing compression rate the area specific felt resistance drops from 0.15 mΩ cm^2 to 0.08 mΩ cm^2 and the relative standard

deviation decreases from 15% to 12% as well, indicating a more homogeneous resistance distribution for high compression rates. In accordance with the effective felt resistance the total cell resistance reduces from 1.24 mΩ cm^2 to 0.98 mΩ cm^2. This reduction of 0.26 mΩ cm^2 is more than three times higher than the diminution of the felt resistance of only 0.07 mΩ cm^2. These two values differ due to the fact, that the cell resistance consists of two compression rate dependent resistances: The felt resistance and the electrolyte resistance. Since both, the felt and electrolyte resistances, lower at higher compression rates, the decline of the cell resistance is more significant than the decrease of the felt resistance on its own.

The averaged dynamic resistances can be used to determine the current densities at the positions of the potential probes. Instead of using external shunt resistors to localize current densities the felt is used as an internal shunt resistor. The current density at the position of each probe can be calculated using Ohm's law and the previously determined averaged dynamic resistance $R_{dyn,i}$.

$$j_i = \frac{\varphi_i}{R_{dyn,i}} \tag{5}$$

Here, j refers to the local current density for probe number i, and φ_i is the measured quasi-stationary probe potential. The polarization curves under quasi stationary conditions for the cell with the highest felt compression rate of 42% are shown in Fig. 8. Besides the current density averaged over the full cell, the local current densities for three potential probes distributed evenly over the height on the right-hand side of the cell (probe 1 close to the inlet, 2 centered in height, 3 close to the outlet) and for the next three probes (4–6) symmetrically positioned on the left-hand side of the cell are depicted. It should be noted, that a higher slope of the

M. Becker et al. / Journal of Power Sources 307 (2016) 826–833

quasi-stationary signal of a potential probe (i.e. Fig. 4 B) leads to a higher current density, thus a lower slope in a plot like Fig. 8.

The pairs of potential probes that are positioned at the same height within the cell (1 and 4, 2 and 5, 3 and 6) show very comparable local current densities for high and low flow rates as well as high or low SoC (Fig. 8). An inhomogeneous flow distribution, e.g. higher flow rates at the left and lower at the right side of the cell due to a blocked or insufficiently dimensioned supply channel, would probably lead to increased current densities at the left side of the cell. Since this is not the case we assume a homogeneous flow distribution within the cell independent of flow rate and SoC. Nevertheless, the flow rate has a distinct effect on the current density distribution along the height of the cell. At low flow rates of only ca. 56 mL/min (Fig. 8 A) the current density in the inlet region of the felt is always higher than the averaged current density for the full cell, while the one in the center or close to the outlet is always lower. As expected, this is the result of high electrolyte utilization leading to fully discharged electrolyte during discharging at low SoC and fully charged electrolyte during charging at high SoC. A much more homogeneous current density is achieved when using high flow rates of 491 mL/min (Figs. 8C and D). Even at low SoCs of 14–18% and a cell voltage of 0.6 V the minimum current density at the cell outlet still achieves nearly 80% of the nominal current density of 4 kA/m^2, while it reaches only 32% at a flow rate of 56 mL/min and a nominal current density of 1.5 kA/m^2.

4. Conclusion

In the present work potential probes were applied to a vertical flow through vanadium redox-flow battery during polarization curve measurements. Due to alternating of the cell voltage, charge and discharge currents occurred. The separation in dynamic and quasi-stationary data allowed the determination of the electrical resistance distribution of the carbon felt. The effective resistances were found to be in the range of 0.08 mΩ cm^2 to 0.15 mΩ cm^2 at compression rates between 9 and 42%. The effective felt resistance at low compression rates showed a dependence on flow rate at the electrolyte inlet that might be attributed to fragile fiber—fiber interconnections that are isolated at high and turbulent flow rates.

Further analysis used the localized felt resistances as internal shunt resistances to calculate the current density distribution under quasi stationary conditions. It appears that the current density is evenly distributed perpendicular to the main flow direction and has a gradient along the main flow direction that is strongly depending on flow rate. Overall, it was successfully demonstrated that potential probes can be used to determine current density distributions in a redox-flow cell. These potential probes are inexpensive, can be easily produced and may be employed in large cells. Its application to cell designs using flow fields will be the issue of future work.

Acknowledgment

The authors gratefully acknowledge ThyssenKrupp Industrial Solutions AG for financial support and the Energy Research Center of Lower Saxony (EFZN) for supporting this project.

References

[1] M. Skyllas-Kazacos, G. Kazacos, G. Poon, H. Verseema, Int. J. Energy Res. 34 (2010) 182–189.
[2] M. Skyllas-Kazacos, M.H. Chakrabarti, S.A. Hajimolana, F.S. Mjalli, M. Saleem, J. Electrochem. Soc. 158 (2011) R55–R79.
[3] Q. Zheng, X. Li, Y. Cheng, G. Ning, F. Xing, H. Zhang, Appl. Energy 132 (2014) 254–266.
[4] J. Friedl, C.M. Bauer, A. Rinaldi, U. Stimming, Carbon 63 (2013) 228–239.
[5] J.-Z. Chen, W.-Y. Liao, W.-Y. Hsieh, C.-C. Hsu, Y.-S. Chen, J. Power Sources 274 (2015) 894–898.
[6] A.M. Pezeshki, J.T. Clement, G.M. Veith, T.A. Zawodzinski, M.M. Mench, J. Power Sources 294 (2015) 333–338.
[7] C. Jia, Y. Cheng, X. Ling, G. Wei, J. Liu, C. Yan, Electrochim. Acta 153 (2015) 44–48.
[8] C.-H. Lin, M.-C. Yang, H. J. Wei, J. Power Sources 282 (2015) 562–571.
[9] S. Liu, L. Wang, B. Zhang, B. Liu, J. Wang, Y. Song, J. Mater. Chem. A 3 (2015) 2072–2081.
[10] J. Xi, Z. Li, L. Yu, B. Yin, L. Wang, L. Liu, X. Qiu, L. Chen, J. Power Sources 285 (2015) 195–204.
[11] B. Yin, Z. Li, W. Dai, L. Wang, L. Yu, J. Xi, J. Power Sources 285 (2015) 109–118.
[12] B. Zhang, E. Zhang, G. Wang, P. Yu, Q. Zhao, F. Yao, J. Power Sources 282 (2015) 328–334.
[13] S. Kim, M. Vijayakumar, W. Wang, J. Zhang, B. Chen, Z. Nie, F. Chen, J. Hu, L. Li, Z. Yang, Phys. Chem. Chem. Phys. 13 (2011) 18186–18193.
[14] L. Li, S. Kim, W. Wang, M. Vijayakumar, Z. Nie, B. Chen, J. Zhang, G. Xia, J. Hu, G. Graff, J. Liu, Z. Yang, Adv. Energy Mater. 1 (2011) 394–400.
[15] M. Vijayakumar, L. Li, G. Graff, J. Liu, H. Zhang, Z. Yang, J.Z. Hu, J. Power Sources 196 (2011) 3669–3672.
[16] M. Vijayakumar, W. Wang, Z. Nie, V. Sprenkle, J. Hu, J. Power Sources 241 (2013) 173–177.
[17] X.K. Ma, H.M. Zhang, C.X. Sun, Y. Zou, T. Zhang, J. Power Sources 203 (2012) 153–158.
[18] A. Tang, J. Bao, M. Skyllas-Kazacos, J. Power Sources 248 (2014) 154–162.
[19] S. Rudolph, U. Schroeder, R.I. Bayanov, K. Blenke, I.M. Bayanov, J. Electroanal. Chem. 736 (2015) 117–126.
[20] Q. Xu, T.S. Zhao, P.K. Leung, Appl. Energy 105 (2013) 47–56.
[21] T. Jyothi Latha, S. Jayanti, J. Power Sources 248 (2014) 140–146.
[22] X. Ke, J.I.D. Alexander, J.M. Prahl, R.F. Savinell, J. Power Sources 270 (2014) 646–657.
[23] C. Yin, Y. Gao, S. Guo, H. Tang, Energy 74 (2014) 886–895.
[24] X.Y. Ke, J.I.D. Alexander, J.M. Prahl, R.F. Savinell, J. Power Sources 288 (2015) 308–313.
[25] T.J. Latha, S. Jayanti, J. Appl. Electrochem. 44 (2014) 995–1006.
[26] S.J.C. Cleghorn, C.R. Derouin, M.S. Wilson, S. Gottesfeld, J. Appl. Electrochem. 28 (1998) 663–672.
[27] J. Stumper, S.A. Campbell, D.P. Wilkinson, M.C. Johnson, M. Davis, Electrochim. Acta 43 (1998) 3773–3783.
[28] W.-Y. Hsieh, C.-H. Leu, C.-H. Wu, Y.-S. Chen, J. Power Sources 271 (2014) 245–251.
[29] J.T. Clement, T.A. Zawodzinski, M.M. Mench, ECS Trans. 58 (2014) 9–16.
[30] D. Aaron, C.-N. Sun, M. Bright, A.B. Papandrew, M.M. Mench, T.A. Zawodzinski, ECS Electrochem. Lett. 2 (2013) A29–A31.
[31] E. Ventosa, M. Skoumal, F.J. Vázquez, C. Flox, J.R. Morante, J. Power Sources 271 (2014) 556–560.
[32] Q. Liu, A. Turhan, T.A. Zawodzinski, M.M. Mench, Chem. Commun. 49 (2013) 6292–6294.
[33] M.R. Mohamed, H. Ahmad, M.N.A. Seman, S. Razali, M.S. Najib, J. Power Sources 239 (2013) 284–293.
[34] S.S.-K. Corcuera, Maria, Eur. Chem. Bull. 1 (2012) 511–519.
[35] W.D. Treadwell, R. Nieriker, Helv. Chim. Acta 24 (1941) 1098–1105.
[36] X.-G. Yang, Q. Ye, P. Cheng, T.S. Zhao, Appl. Energy 145 (2015) 306–319.
[37] R.M. Darling, A.Z. Weber, M.C. Tucker, M.L. Perry, J. Electrochem. Soc. 163 (2016) A5014–A5022.

4 Kinetic studies at carbon felt electrodes for vanadium redox-flow batteries under controlled transfer current density conditions

Electrochimica Acta 252 (2017) 12–24

Contents lists available at ScienceDirect

Electrochimica Acta

journal homepage: www.elsevier.com/locate/electacta

Kinetic studies at carbon felt electrodes for vanadium redox-flow batteries under controlled transfer current density conditions

Maik Becker[a,b,*], Niels Bredemeyer[c], Nils Tenhumberg[c], Thomas Turek[a,b]

[a] Clausthal University of Technology, Institute of Chemical and Electrochemical Process Engineering, Leibnizstr. 17, 38678 Clausthal-Zellerfeld, Germany
[b] Energy Research Centre of Lower Saxony (EFZN), Am Stollen 19A, 38640 Goslar, Germany
[c] thyssenkrupp Industrial Solutions AG, Friedrich-Uhde-Str. 15, 44141 Dortmund, Germany

ARTICLE INFO

Article history:
Received 22 March 2016
Received in revised form 9 July 2017
Accepted 10 July 2017
Available online 21 July 2017

Keywords:
vanadium redox-flow battery
electrode kinetics
mathematical model
current density distribution
charge transfer coefficient

ABSTRACT

In this work we present a combination of mathematical modeling and experimental kinetic characterization of carbon felt electrodes in positive electrolyte (PE) and negative electrolyte (NE) of vanadium redox-flow batteries. The mathematical model is applied to check for homogeneous transfer current density within a radially connected carbon electrode intensively flown through by the electrolyte. Transfer current homogeneity depends mainly on electrolyte conductivity, electrode thickness and charge transfer coefficient. The transfer current inhomogeneity of the investigated electrode samples is below 5%, when single layers of carbon electrodes (415 μm thickness at 89% porosity and 210 μm thickness at 75% porosity) are applied and high electrolyte conductivities ($>800\,\mathrm{mS\cdot cm^{-1}}$) are maintained by low vanadium concentrations ($<15\,\mathrm{mM}$). Experimental results show charge transfer coefficients for high overpotentials of 0.26 ± 0.03 for the reduction reaction in NE and 0.37 ± 0.04 for the oxidation reaction in NE. The charge transfer coefficients of the PE are 0.13 ± 0.02 for reduction and 0.30 ± 0.04 for the oxidation reaction. Application of the Butler-Volmer equation to describe the polarization behavior shows adequate agreement with the experimental results at states of charge between 25% and 75%. The rate constants for the PE reaction are nearly two times higher than for the NE reaction.

© 2017 Elsevier Ltd. All rights reserved.

1. Introduction

Vanadium redox-flow batteries (VRFB) might be an attractive energy storage technology, due to their long cycle life and high energy efficiencies. Further development is vital to increase the power density of the cells and to yield lower power specific costs of the system. Development in the area of cell design and cell setup deals with the improvement of the three main components: the bipolar plate, the separator and the carbon felt.

The development of thin "zero-gap" cells lead to dramatic gain of the achievable power density of VRFBs, since they reduce ohmic resistance within the porous electrode significantly [1]. But the application of single layer carbon paper electrodes, instead of carbon felt electrodes, reduces the surface area as well. Thus, an optimum between thin electrodes with low ohmic potential drop and low surface area leading to high overpotential and thick electrodes with high ohmic potential drop but high surface area that enables low overpotential [2,3] must be found. The optimized thickness is depending on the porosity, electrical conductivity and kinetic behavior of the carbon paper, giving rise to a need for precise information about the kinetic characteristics of the carbon paper.

The modeling of VRFBs is another important field, which is dependent on accurate kinetic data of carbon felt electrodes. Several models use assumed charge transfer coefficients α = 0.5 for both half cell reactions [4–11]. Knehr et al. adjusted the coefficients in order to describe experimental results, but the obtained charge transfer coefficients still remained close to 0.5 [12]. This is in contrast to published data, especially for the positive half cell reaction, as can be seen from Table 2.

First characterization of the kinetics of the V^{2+}/V^{3+} reaction (abbreviated as NE reaction) and the V^{4+}/V^{5+} reaction (abbreviated as PE reaction) were performed by Skyllas-Kazacos group [13,14] using cyclic voltammetry with a stationary disc electrode for both reactions and a rotating disc electrode for the PE reaction only. These authors used glassy carbon as well as gold and platinum electrodes. Since only the glassy carbon electrode showed peak currents and because of its wider potential window without

* Corresponding author at: Clausthal University of Technology, Institute of Chemical and Electrochemical Process Engineering, Leibnizstr. 17, 38678 Clausthal-Zellerfeld, Germany.
E-mail address: maik.becker@tu-clausthal.de (M. Becker).

http://dx.doi.org/10.1016/j.electacta.2017.07.062

M. Becker et al./Electrochimica Acta 252 (2017) 12–24

hydrogen or oxygen evolution, carbon electrodes were the recommended choice for vanadium redox-flow batteries. The charge transfer coefficient determined for the NE reaction was found to be 0.55 but had a large uncertainty (± 0.32). Further cyclic voltammetric studies were published by Kaneko et al. [15], who reported charge transfer coefficients of 0.30 to 0.35 for the reduction and 0.52 to 0.59 for the oxidation of the NE reaction. Instead of using cyclic voltammetry Fabjan et al. [16] performed polarization curve measurements on a smooth surface composite electrode in a usual three electrode assembly under vigorous stirring for sufficient mass transport. Oriji et al. [17] applied potential step experiments on a glassy carbon electrode to overcome mass transport resistances, but their results allowed determining charge transfer coefficients for the NE reaction only. Gattrell et al. [18,19] studied the PE reaction at a rotating disk electrode and reported very high Tafel slopes of 350–450 mV per decade (corresponding to values of 0.13 to 0.17 for the charge transfer coefficient) for the V^{5+} reduction at high overpotentials. Aaron et al. measured the NE kinetics in situ [20] by applying a dynamic hydrogen reference electrode within a VRFB, that operated with an diluted electrolyte of only 0.1 M vanadium in 5 M sulfuric acid. The high exchange current density for the PE reaction lead to small overpotentials for the PE electrode. Therefore, no linear behavior in the Tafel plot could be observed and no charge transfer coefficient could be estimated for this reaction. Agar et al. [21] performed cyclic voltammetry on untreated and heat treated carbon felt electrodes. Since no reduction peak current density was obtained for the NE reaction at untreated electrodes, no evaluation for the NE reduction was possible. Wang et al. [22] performed polarization curves on glassy carbon and pyrolytic graphite rotating disk electrodes to investigate the oxidation of V^{4+}. Additionally they investigated the reduction of V^{5+} [23] and found charge transfer coefficients for cathodic and anodic PE reaction that were comparable to the results published by Gattrell et al. [18,19].

The resulting charge transfer coefficients for the NE and PE reaction are summarized in Tables 1 and 2. The data show a broad variation, probably due to the different methods and different carbon materials applied for characterization. Characterization of porous carbon felt and carbon paper electrodes would be advantageous instead of planar glassy carbon or pyrolytic graphite electrodes, since only the porous electrodes are applied in VFRBs. However, Goulet et al. [24] argued that typical characterization methods as cyclic voltammetry or Tafel analysis on rotating disk electrodes suffer from diffusion limitations if porous electrodes are employed. In situ methods might overcome this problem, but asymmetric exchange current densities impede sufficiently high overpotentials to determine Tafel slopes for the PE reaction. Goulet et al. [24] published a method for direct measurements of the

reaction kinetics on porous electrodes utilizing a flow through setup. They performed experiments with a Toray carbon paper electrode placed within a microfluidic electrochemical cell. Flow velocities as high as 0.1 m·s^{-1} should enhance the mass transfer, so that only kinetic effects would dominate the charge transfer. This technique offers the possibility to determine exchange current densities and charge transfer coefficients at constant operation conditions (state of charge, flow rate) for both NE and PE reaction. However, the Tafel slopes for the PE reaction did not reach a constant value independent of the flow rate. Therefore Goulet et al. [24] stated that further development is required to quantify the influence of the electrode dimensions, which might cause edge effects.

In this work, we present a model to describe the liquid and solid phase potential distribution as well as the transfer current density distribution within a porous electrode. The model is used to quantify the effects of different kinetic parameters, electrolyte conductivities, electrode resistivities and dimensions on the homogeneity of transfer current density. The results are applied to an experimental setup and two different carbon electrodes are characterized at different states of charge (SoC). The kinetic parameters of these materials are then determined at carefully controlled conditions.

2. Experimental

2.1. Materials

The investigated carbon electrodes were SIGRACET® GDL10AA carbon paper (SGL Carbon GmbH, Germany) and Freudenberg H2315 carbon felt (Freudenberg FCCT KG, Germany), which were used as received without any pretreatment. The initial electrolyte was a 1:1 mixture of V^{3+} and V^{4+} supported by GfE (Gesellschaft für Elektrometallurgie, Germany) with a total vanadium concentration of 1.6 M and a total sulfate concentration of 4 M. The electrolyte was charged within a conventional redox flow battery to a SoC of approximately 95%. The precise SoC of charged negative electrolyte

Table 2
Published charge transfer coefficients for the PE reaction.

	α_{cat}	α_{an}	Electrode	Method	Ref
PE	0.33–0.39		composite electrode	polarization curve	[16]
PE	0.13	0.42	glassy carbon	polarization curve	[18,19]
PE	0.23	0.23	raw carbon felt	cyclic voltammetry	[21]
PE	0.14	0.13	heat treated carbon felt	cyclic voltammetry	[21]
PE		0.53	graphite	polarization curve	[22]
PE	<0.15		pyrolytic graphite	polarization curve	[23]
PE	>0.14	0.32	carbon paper	polarization curve	[24]

Table 1
Published charge transfer coefficients for the NE reaction.

	α_{cat}[a]	α_{an}[b]	Electrode	Method	Ref
NE	0.55 ± 0.32	$1 - \alpha_{cat}$[a]	glassy carbon	cyclic voltammetry	[14]
NE	0.30–0.35	0.52–0.59	GRC[c]	cyclic voltammetry	[15]
NE	0.4	0.26	composite electrode	polarization curve	[16]
NE	0.50	0.43	glassy carbon	potential step	[17]
NE	0.30	0.29	carbon paper	full cell polarization[d]	[20]
NE		0.21	raw carbon felt	cyclic voltammetry	[21]
NE	0.26	0.31	heat treated carbon felt	cyclic voltammetry	[21]
NE	0.54	0.49	carbon paper	polarization curve	[24]

[a] α_{cat} – cathodic charge transfer coefficient.
[b] α_{an} – anodic charge transfer coefficient.
[c] GRC – graphite reinforced carbon.
[d] full cell polarization – polarization curves recorded in a working VRFB using dynamic hydrogen reference electrode to examine kinetic behavior.

14   M. Becker et al./Electrochimica Acta 252 (2017) 12–24

(NE) and positive electrolyte (PE) was determined via permanga-nometric titration as reported in our previous paper [25]. The total vanadium concentration for the electrochemical characterization was decreased to 5 mM for the PE and 15 mM for the NE by dilution with 4 M sulfuric acid in order to maintain a constant total sulfate concentration during the measurements. The 4 M sulfuric acid was prepared from deionized water (Milli-Q, 18.2 MΩ·cm) and concentrated sulfuric acid (Ph. Eur., Carl-Roth, Germany).

2.2. Carbon electrode characterization

The in plane resistivity ρ_S of the carbon materials was measured with a resistivity meter equipped with a 4-terminal probe (Loresta-GP MCP-T610 resistivity meter and a PSP probe with 1.5 mm electrode spacing, Mitsubishi Chemical Analytech, Japan). In an evenly distributed pattern on the electrode raw material (1250 cm^2 sheet of H2315 and 625 cm^2 sheet of 10AA material) at least 200 values were measured and averaged. The electrode thickness h and the carbon fiber diameter d_f was measured with a digital microscope VHD-2000D (Keyence, Japan). The carbon density ρ_{carbon} was characterized by a pycnometer Pycnomatic ATC (Porotec, Germany) and was used to calculate the porosity ε of the uncompressed electrode

$$\varepsilon = 1 - \frac{w}{h \cdot \rho_{carbon}} \qquad (1)$$

where w is the areal weight of the electrode material. The specific surface area a of the Freudenberg carbon felt was estimated using the following relationship [26] assuming perfectly round shaped solid carbon fibers without porosity.

$$a = \frac{4}{d_f}(1 - \varepsilon) \qquad (2)$$

Since carbon paper exhibits fiber surface and additional surface from graphite particles, equation (2) could not be applied. Thus the specific surface area of the SGL carbon paper was taken from reference [27].

2.3. Experimental setup

The flow through cell (Fig. 1) consisted of a channel with a constant diameter of 8 mm. Electrode samples with 14 mm diameter were placed into the channel and used as working electrodes. They were contacted to a graphite-polymer compound plate PPG86 (Eisenhuth, Germany) and were sealed with silicone gaskets. An acrylic resin served as insulator to prevent contact of the electrolyte with the graphite plate. The counter electrode was built in a similar way, but instead of a thin (below 0.42 mm thickness) electrode a thick carbon felt (SIGRACELL® GFD4.6EA, SGL, 4.6 mm thickness) electrode was utilized to provide sufficient surface area. A Luggin capillary made of a PTFE tube (1.6 mm outer diameter) between working and counter electrode lead to a hydrogen reference electrode HydroFlex® (Gaskatel, Germany) within 1 M sulfuric acid solution. All potentials provided are referred to this electrode potential, which is close to the standard hydrogen electrode (SHE) potential. To avoid screening of the working electrode by the Luggin capillary, the opening of the capillary had a distance of at least 6 mm to the working electrode. The electrolyte was intensively pumped through this channel by a diaphragm laboratory pump (Liquiport NF 300, KNF, Germany) with a maximum flow rate of 2 L·min^{-1} (corresponding to a flow velocity of 0.66 m·s^{-1}). The electrolyte was supplied by a glass vessel that was purged with pure nitrogen to prevent contamination with atmospheric oxygen. The electrolyte was recirculated between the glass vessel and the flow through cell. Since the redox reaction on working and counter electrode runs in opposite

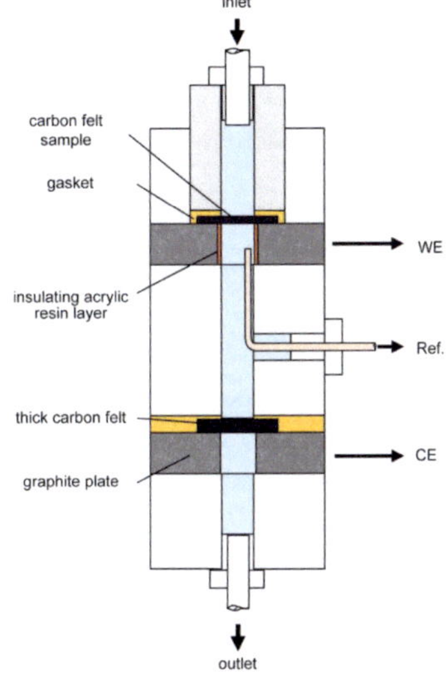

Fig. 1. Experimental flow through setup for 0.5 cm^2 electrode samples (WE: connection of the working electrode, Ref.: connection to the reference electrode, CE: connection of the counter electrode).

directions with exactly the same extent, the total SoC of the electrolyte does not change during an electrochemical characterization. The initially high SoC of nearly 95% was gradually decreased to values of 90%, 75%, 50%, 25%, 10% and 5% by addition of GfE electrolyte with an initial V^{3+}/V^{4+} ratio of 1:1. Constant values of 5 mM or 15 mM total vanadium concentration of PE and NE, respectively, were maintained by addition of 4 M sulfuric acid. The addition of precise volumes of different solutions was ensured by the use of Eppendorf pipettes.

2.4. Electrochemical characterization

A Reference 3000™ potentiostat (Gamry Instruments, USA) was employed for electrochemical open circuit potential measurements, impedance spectroscopy (EIS) and for polarization curve measurements. After measuring the open circuit potential for 60 s, impedance spectroscopy was carried out at the open circuit potential with an a.c. amplitude of 10 mV in a frequency range from 100 kHz to 1 Hz. The polarization curve scan started at −0.25 V to switching potentials of 0.1 V and −0.6 V for the NE reaction. The starting potential for the PE reaction was 1.0 V with switching potentials of 1.5 V and 0.5 V. The scan rate for both reactions was set to 25 mV·s^{-1}. Two cycles with forward and backward scan were performed for both reactions and the resulting data was post-run corrected for iR-drop with the resistance measured at 10 kHz with EIS. To correct the current density from capacitive double layer charging/discharging currents, the current densities of forward

M. Becker et al. / Electrochimica Acta 252 (2017) 12–24

and backward scan at similar potentials were averaged (for further information, see Fig. S1 in the supplementary information). All experiments were performed under room temperature conditions ($21\,°C \pm 2\,°C$).

3. Model of the porous electrode

The mathematical model used in this work is mainly based on a simplified version of the work of Shah et al. [4]. Only slight deviations have to be considered since the electrode under investigation is of cylindrical shape. Furthermore, it is assumed that no mass transport limitations occur, because the flow of electrolyte is so intense that the convective mass transport outperforms all other mass transport effects and leads to very thin diffusion layer thicknesses. Hence no difference between the surface concentration on the carbon fiber and the bulk concentration occurs. Therefore, the surface concentration in each area of the carbon felt equals the concentration within the electrolyte vessel. This assumption simplifies the required equations and only potential and current distribution for the solid and liquid phase have to be combined with a kinetic equation to describe the electrode behavior. A two dimensional model describing the cross section in x-dimension and y-dimension as depicted in Fig. 2 is sufficient due two symmetric conditions. The model is split up in two areas: (I) the porous electrode area and (II) the electrolyte area in the direction of the counter electrode (Fig. 2). The set of equations of the mathematical model is solved by using gPROMS® ModelBuilder 4.0.0 by PSE (Process Systems Enterprise Ltd., UK).

3.1. Equations for the porous electrode

The charge transfer current density i, which corresponds to the faradaic current on the surface of carbon particles and carbon fibers in the carbon electrode, can be determined by the Butler-Volmer equation (3) which is a function of the overpotential η and the exchange current density i_0. The anodic and cathodic charge transfer coefficients are represented by α_{an} and α_{cat}, while F, R and T are Faraday constant, ideal gas constant and temperature,

respectively.

$$i = i_0 \cdot \left(\exp\left(\alpha_{an} \frac{F}{RT} \eta \right) - \exp\left(-\alpha_{cat} \frac{F}{RT} \eta \right) \right) \tag{3}$$

The overpotential η can be described by the difference of solid phase (carbon felt) potential φ_S, the liquid phase (electrolyte) potential φ_L and the open circuit potential E_{eq} (4), which is assumed to follow the Nernst equation (5) taking the concentration of the oxidized c_{ox} or reduced c_{red} vanadium species and the formal redox potential E_{ref} of the reaction under investigation into account.

$$\eta = \varphi_S - \varphi_L - E_{eq} \tag{4}$$

$$E_{eq} = E_{ref} + \frac{RT}{F} \ln\left(\frac{c_{ox}}{c_{red}} \right) \tag{5}$$

The exchange current density i_0 is described by equation (6), where k_0 has the meaning of a reaction rate constant and c_{ref} is the standard concentration of 1 M.

$$i_0 = Fk_0 \cdot c_{ref} \cdot \left(\frac{c_{red}}{c_{ref}} \right)^{\alpha_{cat}} \cdot \left(\frac{c_{ox}}{c_{ref}} \right)^{\alpha_{an}} \tag{6}$$

The charge transfer current density leads to changes of the solid and liquid phase potential affected by the electric resistance of the carbon felt ρ_S, the effective conductivity of the electrolyte κ_{eff} and the specific surface area a of the electrode (eq. (7) and (8)). The effective conductivity of the electrolyte is estimated using the Bruggemann correction (9) and the conductivity of the pure electrolyte, which is assumed to be pure 4 M sulfuric acid [28].

$$\frac{d^2\varphi_S}{dx^2} + \frac{d^2\varphi_S}{dy^2} + \frac{d\varphi_S}{dx}\frac{1}{x} = a \cdot i \cdot \rho_S \tag{7}$$

$$\frac{d^2\varphi_L}{dx^2} + \frac{d^2\varphi_L}{dy^2} + \frac{d\varphi_L}{dx}\frac{1}{x} = -a \cdot i \cdot \frac{1}{\kappa_{eff}} \tag{8}$$

$$\kappa_{eff} = \kappa_L \cdot \varepsilon^{1.5} \tag{9}$$

The resulting current densities of the solid and liquid phase oriented in x and y direction ($j_{S,x}$, $j_{S,y}$, $j_{L,y}$, $j_{S,y}$) can be calculated by Ohm's law.

$$\frac{d\varphi_S}{dx} = -\rho_S \cdot j_{S,x} \tag{10}$$

$$\frac{d\varphi_S}{dy} = -\rho_S \cdot j_{S,y} \tag{11}$$

$$\frac{d\varphi_L}{dx} = -\frac{1}{\kappa_{eff}} \cdot j_{L,x} \tag{12}$$

$$\frac{d\varphi_L}{dy} = -\frac{1}{\kappa_{eff}} \cdot j_{L,y} \tag{13}$$

3.2. Equations for the electrolyte area

The electrolyte area under the carbon electrode can be described in a very similar way, but no solid phase potential and no solid phase current density as well as no charge transfer

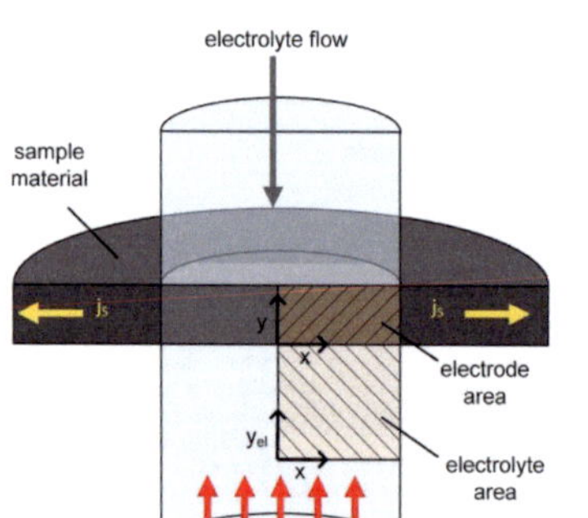

Fig. 2. Modeling region for the carbon felt electrode area and the electrolyte area. The yellow arrows indicate the preferred direction of solid phase current density j_S and the red arrows indicate the liquid phase current density j_L. (For interpretation of the references to color in this figure legend, the reader is referred to the web version of this article.)

    M. Becker et al./Electrochimica Acta 252 (2017) 12–24

current density exists. The liquid phase potential $\varphi_{L,el}$ and the current densities for both orientations $j_{L,y,el}$ and $j_{L,x,el}$ are calculated by equations (14) to (16)

$$\frac{d^2\varphi_{L,el}}{dx^2} + \frac{d^2\varphi_{L,el}}{dy_{el}^2} + \frac{d\varphi_{L,el}}{dx}\frac{1}{x} = 0 \tag{14}$$

$$\frac{d\varphi_{L,el}}{dx} = -\frac{1}{\kappa} j_{L,x,el} \tag{15}$$

$$\frac{d\varphi_{L,el}}{dy_{el}} = -\frac{1}{\kappa} j_{L,y,el} \tag{16}$$

Further variables are required for the calculation of the experimentally available effective current density j_{eff} and the measured potential E_{meas}.

$$j_{eff} = \int_{x=0}^{r} j_{L,x,el}(x,0)dx / \int_{x=0}^{r} x\,dx \tag{17}$$

$$E_{meas} = 0 - \varphi_{L,el}(r,0) \tag{18}$$

The base case parameters required for the mathematical model are listed in Table 4.

3.3. Boundary conditions

The boundary conditions at the center of the electrode prohibit any current density in x-dimension due to symmetry conditions.

$$j_{L,x} = 0 \ (\text{for } x = 0, \ 0 < y < h) \tag{19}$$

$$j_{S,x} = 0 \ (\text{for } x = 0, \ 0 < y < h) \tag{20}$$

At the upper and lower boundary of the electrode there can be no solid phase current density in the y-dimension, since there is no solid phase.

$$j_{S,y} = 0 \ (\text{for } 0 < x < r, \ y = 0) \tag{21}$$

$$j_{S,y} = 0 \ (\text{for } 0 < x < r, \ y = h) \tag{22}$$

Liquid phase current density at the top of the electrode is comparably unlikely, because the counter electrode is disposed below the working electrode.

$$j_{L,y} = 0 \ (\text{for } 0 < x < r, \ y = h) \tag{23}$$

Boundary conditions for the contact area between porous electrode and graphite plate have been considered by assuming constant solid phase potential and no current density in the liquid phase.

$$\varphi_S = 0 \ (\text{for } x = r, \ 0 < y < h) \tag{24}$$

$$j_{L,x} = 0 \ (\text{for } x = r, \ 0 < y < h) \tag{25}$$

Boundary conditions for the intersection between electrode and electrolyte area have to maintain constant potential and constant current density for the liquid phase potential.

$$\varphi_L = \varphi_{L,el} \ (\text{for } 0 < x < r, \ y = 0, \ y_{el} = h_{el}) \tag{26}$$

$$j_{L,y} = j_{L,y,el} \ (\text{for } 0 < x < r, \ y = 0, \ y_{el} = h_{el}) \tag{27}$$

Similar to the porous electrode no current density is possible in x-dimension at the center and at the outer edge of the channel.

$$j_{L,x,el} = 0 \ (\text{for } x = 0, \ 0 < y_{el} < h_{el}) \tag{28}$$

$$j_{L,x,el} = 0 \ (\text{for } x = r, \ 0 < y_{el} < h_{el}) \tag{29}$$

The current density in x-dimension of the electrolyte area not necessarily equals zero but it is highly unlikely that there is a current in x-dimension at the bottom of electrolyte area, since the main current density would be oriented in y-dimension. Therefore, we assumed a constant liquid phase potential in x-dimension, described by a gradient of the liquid phase potential of zero.

$$\frac{d\varphi_L}{dx} = 0 \quad (\text{for } 0 < x < r, \ y_{el} = 0) \tag{30}$$

3.4. Equations used for evaluation of the experiments

If the experimental conditions maintain a homogeneous charge transfer current density, the Butler-Volmer equation can be applied (31). The overpotential is described by the difference of iR-drop corrected potential E and the equilibrium potential, calculated by a simplified Nernst equation only considering the vanadium species (32).

$$j_{eff} = F \cdot a h \cdot k_0 \cdot \left(\frac{c_{red}}{c_{ref}}\right)^{\alpha_{cat}} \cdot \left(\frac{c_{ox}}{c_{ref}}\right)^{\alpha_{an}} \cdot \left(\exp\left(\alpha_{an}\frac{F}{RT}\eta\right) - \exp\left(-\alpha_{cat}\frac{F}{RT}\eta\right)\right) \tag{31}$$

$$\eta = E - E_{ref} - \frac{RT}{F} \ln\left(\frac{c_{ox}}{c_{red}}\right) \tag{32}$$

The calculation of charge transfer coefficients bases on the Tafel equation

$$\eta = A + B \cdot \log_{10}|j_{eff}| \tag{33}$$

where A and B are constants and B, also called Tafel slope, allows the calculation of the charge transfer coefficients [29].

$$\alpha = \frac{RT}{BF} 2.3 \tag{34}$$

4. Results and Discussion

4.1. Carbon felt characterization

The microscopic characterization of Freudenberg H2315 carbon felt electrode lead to an average carbon fiber thickness of 10.7 ± 1.1 μm. The fiber density of the carbon felt was determined to 1.92 g·cm^{-3} in comparison to the bulk density of 0.49 g·cm^{-3} which results in a porosity of 75% and a specific surface area of

Table 3
Carbon electrode properties of Freudenberg H2315 carbon felt and SGL SIGRACET® 10AA GDL carbon paper.

Material	Parameter	Value	Unit	Ref.
H2315	a	$9.4 \cdot 10^4$	m^2·m^{-3}	This work
H2315	ε	0.75	–	This work
H2315	h	$0.21 \cdot 10^{-3}$	m	This work
H2315	ρ_s	$0.168 \cdot 10^{-3}$	Ω·m	This work
10AA	a	$60 \cdot 10^4$	m^2·m^{-3}	[27]
10AA	ε	0.89	–	This work
10AA	h	$0.415 \cdot 10^{-3}$	m	This work
10AA	ρ_s	$0.198 \cdot 10^{-3}$	Ω·m	This work

$9.4 \cdot 10^4 \, m^2 \cdot m^{-3}$. The carbon paper supplied by SGL exhibits a significantly higher porosity of 89%, although the carbon density is nearly the same. The rounded value of the electrical resistivities of both electrodes was $0.2 \cdot 10^{-3} \, \Omega \cdot m$, but the carbon felt resistivity was somewhat lower, which might be attributed to the lower porosity. The precise carbon electrode parameters are listed in Table 3.

4.2. Check for homogeneous charge transfer current density distribution by mathematical modeling

Modeling was carried out to check for a homogeneous charge transfer current density distribution and to investigate the effect of different electrode thicknesses, electrolyte resistivities and kinetic parameters. In general, one could model the charge transfer current density distribution for NE and PE, each, however these results for PE and NE would be identical, if the reaction rate constant, the charge transfer coefficient, the concentrations, electrode dimensions and conductivities would be the same. Therefore, we modeled only the charge transfer current density distribution for the PE electrode under variation of these parameters.

As a first example a polarization curve was modeled for the base case parameters (corresponding to three layers of Freudenberg H2315 carbon felt and most other parameters as stated in Table 4). Only the total vanadium concentration was set to 1.6 M to equal the concentration in commercially available electrolytes. Therefore the electrolyte conductivity was decreased to approximately $35 \, S \cdot m^{-1}$ [30]. To give an overview about the current and potential distribution, Fig. 3 shows the solid and liquid phase potentials, overpotentials and charge transfer current densities for an exemplary effective current density of $1000 \, A \cdot m^{-2}$. A high gradient for the solid phase potential in x-direction and an even higher gradient of the liquid phase potential in the y-direction occurs.

Table 4
Base case parameters unless otherwise stated.

Parameter	Value	Unit	Ref.
k_0	$3 \cdot 10^{-9}$	$m \cdot s^{-1}$	[4]
$\alpha_{cat}, \alpha_{an}$	0.5	–	[4]
E_{ref}	1.005	V	[4]
F	96485	$C \cdot mol^{-1}$	
R	8.314	$J \cdot mol \cdot K^{-1}$	
T	298	K	
c_{ox}, c_{red}	2.5	$mol \cdot m^{-3}$	
c_{ref}	1000	$mol \cdot m^{-3}$	
a	$9.4 \cdot 10^4$	$m^2 \cdot m^{-3}$	chosen as H2315
ε	0.75	–	chosen as H2315
h	$0.63 \cdot 10^{-3}$	m	chosen as 3 layers of H2315
r	$4 \cdot 10^{-3}$	m	
κ_L	82.3	$S \cdot m^{-1}$	[28]
ρ_s	$0.2 \cdot 10^{-3}$	$\Omega \cdot m$	chosen as rounded value of 10AA

Hence, the solid phase current density is mainly oriented in x-direction, while the liquid phase current density is oriented in y-direction. The overpotential results from the combination of liquid and solid phase potential. It exhibits a maximum at the edge close to the graphitic current collector and the counter electrode while a minimum on the diametrically opposed position close to the center of the channel at the top of the electrode exists. The charge transfer current density has a congruent but more pronounced shape due to its exponential dependence on the overpotential. Thus, generally small differences in the overpotential of only 15 mV lead to deviations between maximum charge transfer current density and average current density of nearly 20%.

The modeled polarization curve as shown in the Tafel plot of Fig. 4 exhibits deviations from the ideal polarization curve obtained by the Butler-Volmer equation (31) in the absence of distributions. Although the differences seem to be small in the logarithmic presentation of the Tafel plot, the relative error shows

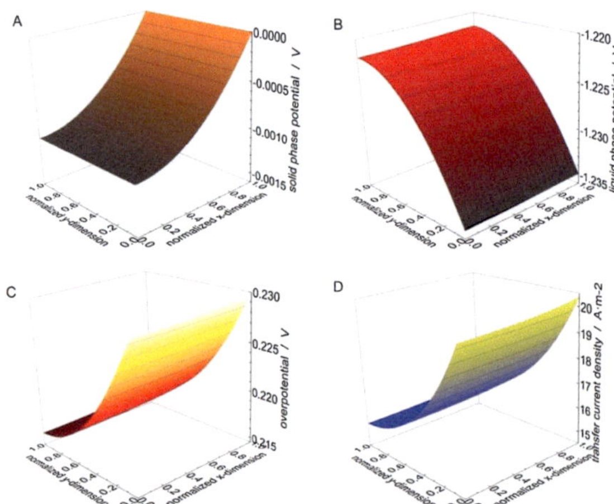

Fig. 3. Solid and liquid phase potential distribution (A and B) as well as overpotential and resulting charge transfer current density (C and D) modeled for 50% SoC and oxidation reaction at an effective current density of $1000 \, A \cdot m^{-2}$.

18   M. Becker et al./Electrochimica Acta 252 (2017) 12–24

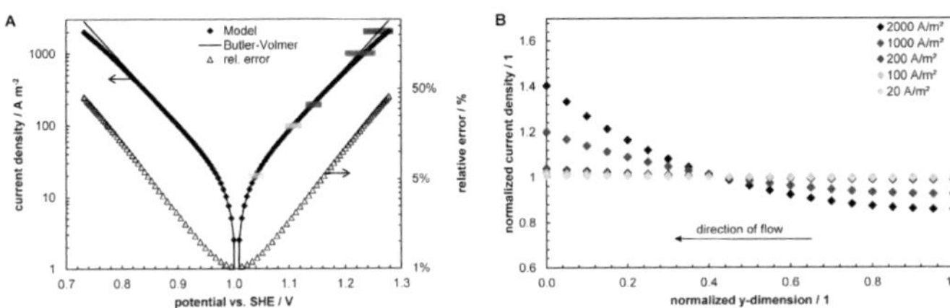

Fig. 4. A) Modeled and calculated polarization curve and the resulting relative error for a SoC of 50%. B) Normalized current density adjacent to the current collector ($x = r$) shown for different applied effective current densities (as indicated by colored bars in the polarization curve).

deviations of almost 40% for an effective current density of 2000 A·m^{-2}. This deviation arises when significant gradients in the overpotential occur, which induce an unevenly distributed charge transfer current density. For quantification of the homogeneity of the charge transfer current density for different current densities applied to the electrode we used a normalized current density, which is defined as the ratio of the local charge transfer current density and the average charge transfer current density of the entire electrode. In Fig. 4 the normalized current density is plotted versus the normalized y-dimension since the changes are more distinctive in y-direction than in x-direction (y-axis value of Fig. 4 is normalized to the carbon electrode height). For the same reason the plot is only depicted for $x = r$, close to the current collector. It is obvious that higher effective current densities lead to higher potential gradients and therefore to higher non-uniformity of the charge transfer current density. The highest normalized charge transfer current density appears for $y = 0$ at every effective current density and its value shows a very good correlation to the relative error between modeled and calculated current densities for the ideally homogeneous case. For an effective current density of 1000 A·m^{-2} the maximum current density is 20% higher than the average current density and this is similar to the relative error between calculated and modeled current density which gives a value of 20%, too. Therefore, we investigated the effects of reaction rate constants, vanadium concentration, electrolyte conductivity, electrode thickness and charge transfer coefficient on the distribution of the normalized current density. To reduce the computing time we fixed the current density to a value of 1000 A·m^{-2} and did not always calculate a full polarization curve, as the deviations from homogeneity increase at higher current densities. If we avoided current densities in the experimental setup higher than this fixed current density, we could use the model to predict maximum deviations from homogeneity and furthermore the model allowed us to reduce these deviations by proper choice of experimental conditions. The base case parameters (Table 4) were applied to the model.

Changing the reaction rate constant by several orders of magnitude or applying different total vanadium concentrations showed no effect on the charge transfer current density distribution, since these values only alter the exchange current density, but they have no influence on the relationship between overpotential and charge transfer current density, as long as no other parameters are changed. However, it is important to note that different total vanadium concentrations lead to different electrolyte conductivities [31] and therefore indirectly affect the charge transfer current density. Furthermore, it should be stated that higher vanadium concentrations or reaction rate constants would decrease the required overpotential, so Tafel slopes might not be experimentally

accessible, since no linear behavior in the Tafel plot would be apparent.

The effect of different electrolyte conductivities is shown in Fig. 5. Higher electrolyte conductivities of 80 S·m^{-1} or 100 S·m^{-1} result in a significant decrease of the maximum charge transfer current density being less than 10% higher than the average current density. Therefore, we decided to use 4 M sulfuric acid and only very low vanadium concentrations (max. 15 mM Vanadium) in our experiments. We assumed that these low vanadium concentrations do not decrease the conductivity of sulfuric acid, which is 82.3 S·m^{-1} at room temperature [28]. For this base case electrolyte conductivity, we varied the electrode thickness and the charge transfer coefficients as reported in Fig. 6. By decreasing the number of electrode layers, the charge transfer current density becomes more evenly distributed and the maximum normalized current density for one electrode layer is only 6% higher than the average current density. It should be noted that the normalized current density for one layer of the electrode is always higher than unity when measured close to the current collector ($x = r$, shown in Fig. 6A) and always lower than unity when measured at the center of the channel ($x = 0$, not shown). This behavior is caused by an increasing solid phase current density in x-direction for thin electrodes, which leads to higher potential gradients in the solid phase. These potential gradients lower the overpotential in the center of the channel and hence, the charge transfer current density in the center of the channel is lower than at the edge to the current collector. Thus, the charge transfer current density has to be checked in x-dimension and y-dimension. Nevertheless, the exemplary effect of different charge transfer coefficients on the

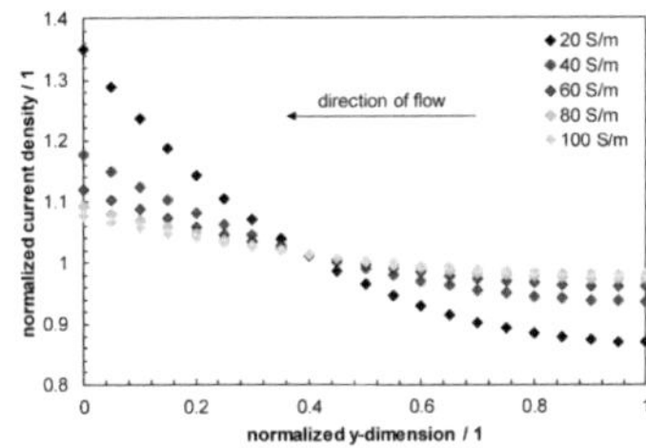

Fig. 5. Effect of different electrolyte conductivities on the normalized charge transfer current density distribution close to the current collector for an effective current density of 1000 A·m^{-2} ($x = r$, base case).

M. Becker et al. / Electrochimica Acta 252 (2017) 12–24

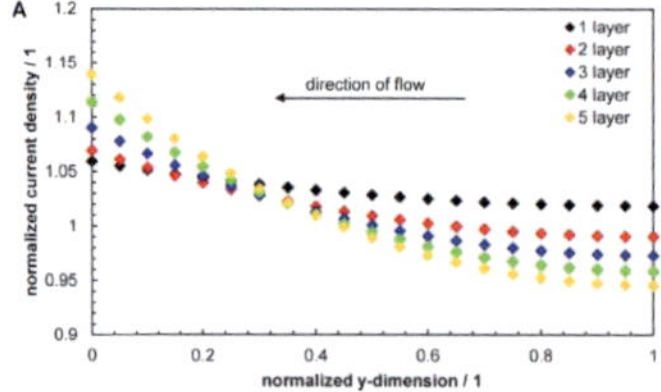
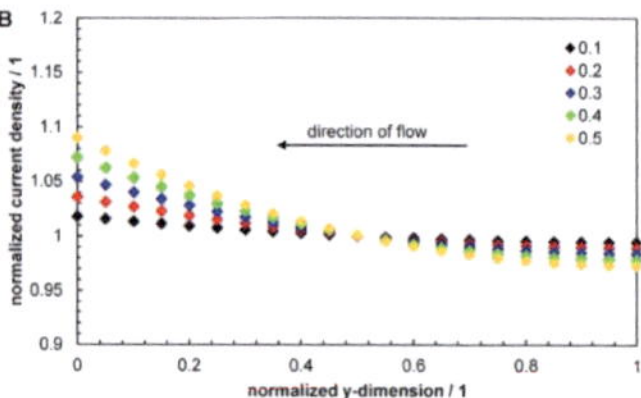

Fig. 6. A) Effect of electrode thickness (thickness of 1 layer is 0.21 mm) and B) different charge transfer coefficients on normalized charge transfer current density distribution close to the current collector for an effective current density of 1000 A·m^{-2} ($x = r$, base case).

normalized charge transfer current density in y-dimension can be seen in Fig. 6B). A lower charge transfer coefficient leads to lower deviations from the average current density, since the charge transfer current density becomes less sensitive to changes in the overpotential.

To quantify the deviations from the average current density for both electrodes under investigation, we used the parameters given in Table 3 and assumed a maximum charge transfer coefficient of 0.4. This assumption is justified, since most of the charge transfer coefficients reported in Tables 1 and 2 are lower or equal than 0.4. Furthermore, this assumption is validated in the results section of our experimental work. We set the electrode height to the height of a single layer electrode and used the electrolyte conductivity of 4 M sulfuric acid. To account for gradients in x- and y-dimension, we report the ratio of the maximum charge transfer current density of the entire electrode to the average charge transfer current density. For the H2315 electrode the maximum charge transfer current density at 1000 A·m^{-2} is 4.4% higher than the average charge transfer current density, while for the 10AA electrode the deviation is 4.6%. Although the 10AA electrode is nearly twice as thick as the H2315 electrode, the deviation is only slightly higher, which can be explained by the higher porosity that leads to higher effective conductivity of the liquid phase and therefore to lower potential gradients in y-dimension. For applied current densities of 2000 A·m^{-2} the resulting deviation still remains below 10% for both electrodes and rises to values of 12% and 14% at a current density of 3000 A·m^{-2} for H2315 and 10AA electrode, respectively.

4.3. Experimental results of kinetic electrode characterization

Considering the previously described modeling results current densities larger than 1000 A·m^{-2} were avoided to guarantee a homogeneous distribution of the charge transfer current density. It was also shown that low total vanadium concentrations are essential to yield a high electrolyte conductivity. Therefore, a vanadium concentration of 15 mM for the NE and of only 5 mM for the PE was applied. The first experiment was carried out with PE at a SoC of 5% which was pumped through a single layer of a 10AA electrode at different flow rates of 0.2 to 2 L·min^{-1}. The resulting polarization curves are shown in Fig. 7. It can be seen that a flow rate of more than 1 L·min^{-1} does not enhance the current density any further. This flow rate corresponds to a change of the electrolyte's SoC of no more than 0.7% at a maximum current density of 1000 A·m^{-2} and hence the modeling assumptions of a constant surface concentration are satisfied. To maintain a sufficient supply of reacting vanadium species for all possible circumstances a flow rate of 2 L·min^{-1} was chosen. Additionally, microscopic images of both electrode samples show a uniform carbon fiber distribution, hence a uniform pore size distribution, so a homogeneous flow velocity in the pores of the electrode is very probable. After recording of the polarization curves the electrode samples were removed from the setup. No structural changes or damages could be observed for all electrodes under investigation. The ohmic resistance of the setup measured by EIS was in the range of 1.48 to 1.62 Ω, which was mainly caused (>90%) by the electrolyte resistance between the tip of the Luggin capillary and the working electrode.

The reproducibility of our method was checked by application of three different electrode samples of each material in the setup. Although the Tafel slopes for the same material did not differ by more than 5%, the resulting current densities exhibited deviations of around 20% for the 10AA material and of less than 8% for H2315. This might be due to variances in electrode thickness, porosity, carbon fiber diameter or carbon surface properties (functional groups, edge or basal plane sites, defect sites).

Further polarization curves were recorded by using a single electrode sample for different SoC of the electrolytes. The results obtained for NE and PE reactions at the 10AA electrode are reported in Fig. 8, where both reactions show a distinctive dependence on the SoC. As expected, the cathodic current increases and the anodic current decreases with rising SoC for the PE reaction. For the NE reaction this dependence is reversed but the cathodic branch of the NE reaction is only slightly influenced by the SoC. This is very probably due to side reactions at low potentials and especially due to the hydrogen evolution reaction. At these low potentials and under these acidic conditions with low V^{3+} concentrations, the hydrogen evolution reaction can become significant, although graphitic materials show a high overpotential for the hydrogen

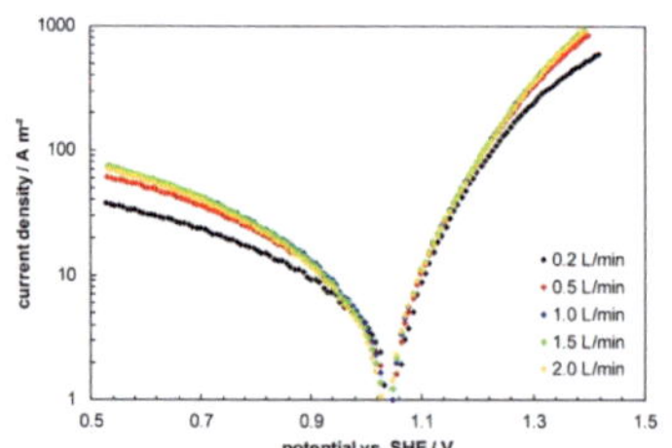

Fig. 7. Effect of different flow rates on the polarization curve of a SGL GDL 10AA carbon paper in PE with 5% SoC.

20   M. Becker et al. / Electrochimica Acta 252 (2017) 12–24

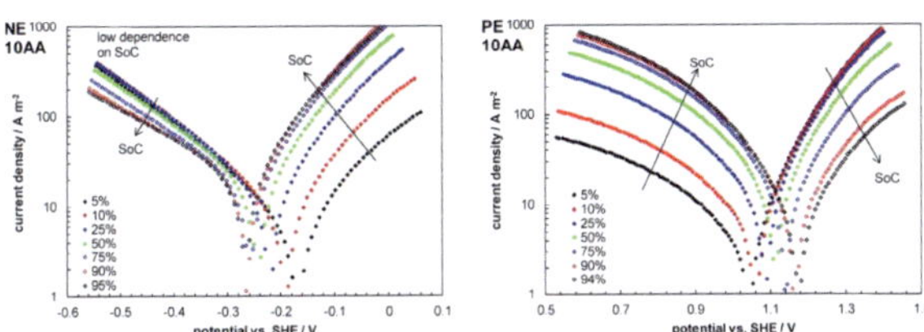

Fig. 8. Polarization curves of SGL GDL 10AA carbon paper for 15 mM NE and 5 mM PE in 4 M H_2SO_4 at different SoC (arrows indicate direction of increasing SoC).

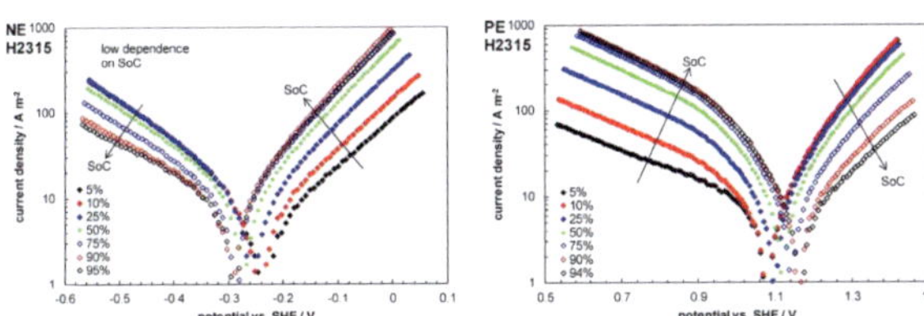

Fig. 9. Polarization curves of Freudenberg H2315 GDL for 15 mM NE and 5 mM PE in 4 M H_2SO_4 at different SoC (arrows indicate direction of increasing SoC).

evolution reaction. This low dependency of the cathodic current density on SoC is apparent for the NE reaction at the H2315 electrode as well (Fig. 9), however it is not as distinctive as for the 10AA material.

For the cathodic branch of the PE reaction at the 10AA electrode, a steady increase of the Tafel slope is observable, instead of an abrupt change as reported by Gattrell et al. [18] or by Wang et al. [23]. In contrast to this behavior, the H2315 electrode shows a rapid change of the Tafel slope at a transition potential of nearly 0.9 V vs. SHE (Fig. 9). Nevertheless, the SoC dependence is comparable to the 10AA electrode.

A first evaluation of the polarization curves was carried out by determining the Tafel slope by linear regression in the Tafel plot. The Tafel slopes of the NE reaction were determined at potentials higher than 100 mV above the open circuit potential for the anodic branch and at potentials lower than 100 mV below the open circuit potential for the cathodic branch. The potential window for determining the Tafel slopes of the PE reaction was adjusted to the open circuit potential as well, but the for the cathodic branch the potential window limit was increased to 200 mV below the open circuit potential to account for the change of the Tafel slope in this branch (the anodic limit remained at 100 mV above the open circuit potential). The resulting charge transfer coefficients presented in Fig. 10 show a good congruence for both electrodes. The charge transfer coefficients for the NE reaction are in sum higher than those of the PE reaction, although the sum does not achieve unity as well. Therefore, we compare our results to published data, where polarization curve measurements were applied on carbon paper electrodes. Our results contradict data from Goulet et al. [24] (cf. Table 1), who used a similar method but

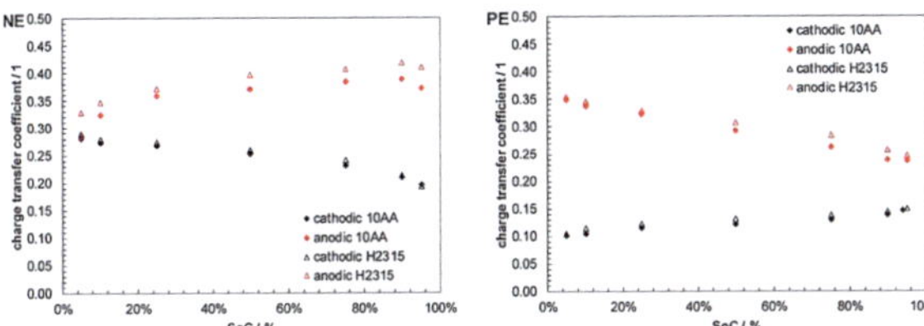

Fig. 10. Anodic and cathodic charge transfer coefficients at high overpotentials of NE and PE reaction for H2315 and 10AA electrodes at different SoC.

M. Becker et al. / Electrochimica Acta 252 (2017) 12–24

significantly higher vanadium concentrations (1.7 M). On the other hand, our results are in good agreement with those from Aaron et al. [20], who used the same material (10AA) and a diluted vanadium concentration of 0.1 M.

If the NE reaction would follow a simple one electron transfer reaction, a charge transfer coefficient in the range of 0.5 would be expected. In contrast, the evaluation of our experimental results gave values of 0.26 and 0.37 for the cathodic and anodic charge transfer coefficients, respectively. Since the Freudenberg H2315 material comprises of solid carbon fibers without any filler material, any kind of inner mass transport limitation that would not be affected by the high flow rate is unlikely. At least for this material, inner mass transport limitations cannot explain the low charge transfer coefficient. On the other hand, mass transport in the 10AA electrode material could be influenced by the carbon filler particles, that are supporting the carbon fibers and build a porous, surface increasing system in the electrode. The pores in this system might not be flown through by the electrolyte, so inner mass transport limitations may occur which could lead to the curvature of the 10AA polarization curves.

The impact of nonuniform charge transfer current density distribution was checked in the modeling section of this paper. Since the charge transfer coefficients are mainly below 0.4 and the applied current density is lower than 1000 A·m^{-2} a very homogeneous distribution is ensured.

The hydrogen evolution side reaction can also influence the cathodic Tafel slopes. However, it would have no influence on the anodic Tafel slope and the resulting anodic charge transfer coefficient, which is significantly below 0.5 as well. Though, it does not explain the deviation from the theoretical expected charge transfer coefficient. The basic difference to the results published by Goulet et al. [24] is the choice of carbon material and the vanadium concentration, which might be the cause for different charge transfer coefficients. Additional measurements for further carbon materials and varying vanadium concentrations

are recommended to verify the correct charge transfer coefficient for the NE reaction. If charge transfer coefficients less than 0.5 can be verified, a more complex reaction mechanism for the NE reaction is probable, which might be due to complexation or adsorption reactions on the carbon surface or due to parallel reactions on different surface sites.

The charge transfer coefficients for the PE reaction are in good agreement with most results published in literature (Table 2), where the very low charge transfer coefficient for the PE reduction reaction is usually explained by a complex multistep reaction mechanism [18].

4.4. Kinetic parameters resulting from fitting calculated to measured data

A second evaluation of the experimental results was exercised by fitting all four parameters of the Butler-Volmer equation (31) (formal redox potential, reaction rate constant, anodic and cathodic charge transfer coefficient) to the experimental data by using the least squares method for the calculated and experimentally measured logarithmized current densities for SoC from 5% to 95%. Since a simple version of the Butler-Volmer equation is unable to handle different Tafel slopes for low and high overpotentials, we limited the description of the PE reaction to potentials higher than 0.9 V vs. SHE. This is of special importance, because the NE reaction is typically more limiting than the PE reaction within an operating VRFB. Therefore, the overpotentials at the PE electrode are lower than at the NE electrode and a good approximation of kinetic parameters in the region of low overpotentials is more useful for mathematical modeling. Fig. 11 presents the comparison of calculated and measured data of both electrode materials for a SoC between 25% and 75%. The polarization curves exhibit a good congruence of measured and calculated data especially for the NE reaction. Deviations arise for the PE polarization curve at high and low overpotentials, since the curvature of the measured data

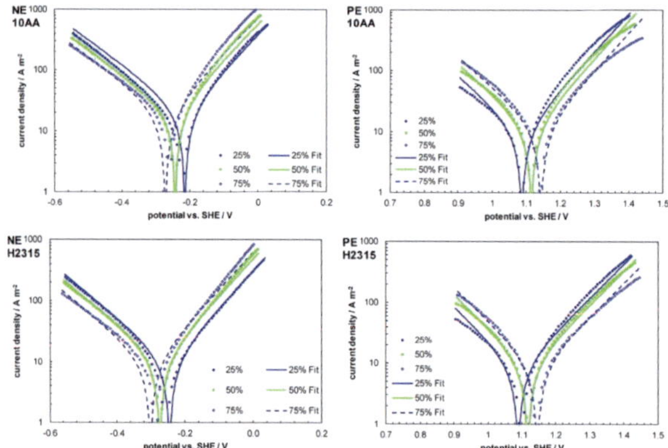

Fig. 11. Polarization curves for the NE and PE reaction at the SGL GDL 10AA material and the Freudenberg H2315 material. Comparison of measured data (dots) and calculated values from the Butler-Volmer equation (lines) for SoC of 25%, 50% and 75% (higher and lower SoCs are not shown for ease of presentation, cf. Fig. 12).

M. Becker et al. / Electrochimica Acta 252 (2017) 12–24

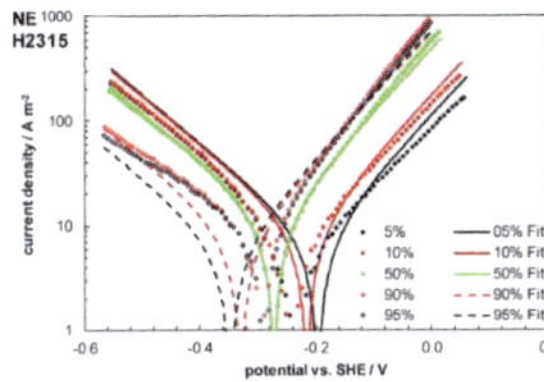
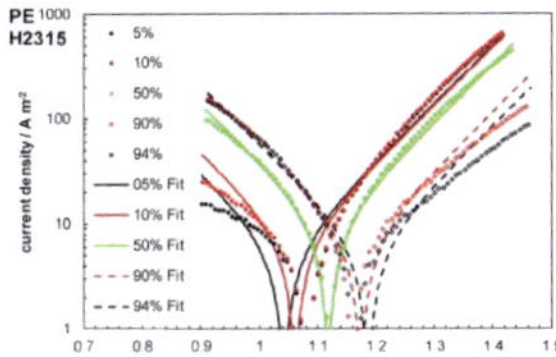

Fig. 12. Polarization curves for the NE and PE reaction at the Freudenberg H2315 material. Comparison of measured data (dots) and calculated values from the Butler-Volmer equation (lines) for SoC of 5%, 10% 50%, 90% and 95% (the comparison between measured and calculated polarization curves for all SoC and both materials can be found in the supplementary information, Fig. S2).

cannot be reproduced with the Butler-Volmer equation. This curvature can probably be attributed to the multi-step reaction mechanism of the PE reaction and since it is more distinctive for the 10AA material, it also might result from remaining mass transport limitations in the porous carbon filler material of the 10AA electrode.

For SoC lower than 25% and higher than 75% the deviations between calculated and measured polarization curves increase significantly as exemplarily shown in Fig. 12 for the H2315 material. This deviation can be caused by a change in the Tafel slope for different SoC as presented in Fig. 10, by the influence of hydrogen evolution side reaction for the cathodic branch in the NE reaction or by a reaction order less than one for both reactions. All of these effects cannot be described by the applied form of the Butler-Volmer equation and would lead to deviations especially for high and low SoC.

The parameters resulting from fitting the calculated polarization curves to the measured polarization curves are listed in Table 5, where similar results for the NE charge transfer coefficients as depicted in Fig. 10 are indicated. Nevertheless the cathodic charge transfer coefficient for the PE reaction is notably shifted upwards to a value between 0.29 and 0.32, which is due to the lower overpotential compared to the charge transfer coefficients shown in Fig. 10. The fitted formal redox potentials for the NE reaction are close to the expected value of −0.255 V vs. SHE [4], but the PE potentials are significantly (ca. 112 mV) higher than the expected value of 1.004 V [4]. This can be attributed to the higher proton activity in 4 M sulfuric acid solution compared to standard conditions and the simplified Nernst equation (32) used in this work that neglects the effect of proton concentration on equilibrium potential. The reaction rate constants show a very noticeable difference of more than one order of magnitude between the H2315 and the 10AA material. This might be caused by the different methods used to determine the specific surface areas and by the unknown true surface area wetted by the electrolyte. The graphite impregnated carbon paper (10AA) could include a higher degree of pores that might not have ionically

conductive access to the electrolyte and the carbon felt (H2315) might have a higher surface area than determined, since surface roughness of the fibers was not experimentally accessible via microscopy. Both effects would lead to more similar true wetted surface areas of both electrodes and would therefore lead to more similar reaction rate constants. Additionally, the surface properties of the carbon material (edge or basal plane sites, defect sites, oxygen functional groups) may cause different reaction rates for the two materials under investigation. To avoid the demand for a precise wetted surface area one could use the exchange current density j_0 as described in equation (35), but its use is limited since it depends on SoC.

$$j_0 = F \cdot ah \cdot k_0 \cdot c_{ref} \cdot \left(\frac{c_{red}}{c_{ref}}\right)^{\alpha_{cat}} \cdot \left(\frac{c_{ox}}{c_{ref}}\right)^{\alpha_{an}} \tag{35}$$

Therefore, we implemented the standard exchange current density j_{00}, which is the exchange current density for standard concentrations of all involved electroactive species [32].

$$j_{00} = F \cdot ah \cdot k_0 \cdot c_{ref} \tag{36}$$

This parameter allows a direct comparison of the kinetic properties of different electrode samples independent from electrode thickness, specific surface area and electrode thickness, if ohmic losses and outer mass transport losses are not limiting the cell setup. Therefore, we recommend the application of this parameter for kinetic investigations of electrode materials. The comparison of both electrodes investigated in this work shows a 30% to 65% higher standard exchange current density for the 10AA electrode compared to the H2315 electrode. Nevertheless, it should be considered that the 10AA electrode is nearly twice as thick as the H2315 electrode and two layers of H2315 would already yield a significantly higher standard exchange current density but still maintain almost the same electrode thickness.

Comparison of NE and PE reaction for each material reveals that the standard exchange current density for the PE reaction is only about two times higher than for the NE reaction. This is a quite low difference compared to results reported in literature [20,24] were higher total vanadium concentrations of 0.1 M or 1.7 M were used. To understand this discrepancy more work has to be done to investigate effects of vanadium concentration, carbon material and electrode pretreatment, falling outside the scope of this work.

5. Conclusions

A mathematical model was used to predict the homogeneity of the charge transfer current density within a porous electrode. The modeling showed that thin electrodes and high electrolyte

Table 5
Kinetic parameters of both investigated carbon electrodes resulting from fitting the parameters to reproduce the measured polarization curves.

Parameter Reaction	H2315 NE	10AA NE	H2315 PE	10AA PE	Unit
E_{ref}	−0.27	−0.24	1.12	1.12	V vs. SHE
k_0	$1.2 \cdot 10^{-7}$	$1.6 \cdot 10^{-8}$	$2.3 \cdot 10^{-7}$	$2.3 \cdot 10^{-8}$	$m \cdot s^{-1}$
j_{00}	232	387	432	563	$A \cdot m^{-2}$
α_c	0.27	0.24	0.32	0.29	
α_a	0.36	0.35	0.33	0.36	

M. Becker et al. / Electrochimica Acta 252 (2017) 12–24

conductivities as well as low charge transfer coefficients are essential for a homogeneous potential and charge transfer current density distribution, while vanadium concentration and reaction rate constant had no effect on the homogeneity for a given effective current density. Since only the electrode thickness and the electrolyte conductivity are adjustable to a certain extent, we investigated single layers of electrode samples in very diluted (<15 mM vanadium) and therefore highly conductive electrolytes. The resulting error due to inhomogeneous charge transfer current density for a given electrode diameter of 8 mm was below 5% for both electrodes, carbon felt and carbon paper material.

An experimental setup with a vertically flown through and radially contacted electrode sample was developed. High flow velocities above $0.33 \, \text{m} \cdot \text{s}^{-1}$ prevented mass transport limitations, thus the pure kinetic behavior of the electrodes was accessible. Polarization curves were recorded for different states of charge during PE and NE reaction and cathodic and anodic Tafel slopes were derived from these measurements. The resulting charge transfer coefficients were in good agreement with the literature for the PE reaction and we could observe a transition potential at the cathodic branch of the PE reaction for the H2315 carbon felt electrode. A slight but steady transition of the Tafel slope was observed for the 10AA carbon paper electrode, that might be caused by inner mass transport limitations, which cannot be avoided even at high electrolyte flow rate.

For the cathodic branch of the NE reaction hydrogen evolution probably influenced the measurement and could affect the cathodic charge transfer coefficient. The charge transfer coefficients for the NE reaction did show only good agreement to results from Fabjan et al., Aaron et al. and Agar et al. [16,20,21] but differed to results from Sum et al., Kaneko et al., Oriji et al. and Goulet et al. [14,15,17,24]. Unfortunately, no clear dependence on carbon material, vanadium or sulfuric acid concentration is recognizable in literature, but generally our results show a better agreement with results of polarization curve measurements than with results from cyclic voltammetry. However, polarization curves are probably the more suitable method to determine Tafel slopes than cyclic voltammetry measurements. In addition to publications of Aaron et al. and Goulet et al. [20,24], we could eliminate or identify effects of mass transport limitation, potential distribution and inhomogeneous charge transfer current density, as well as the dependence of SoC that could influence the measured Tafel slope and the charge transfer coefficient.

The applied Butler-Volmer equation offered an adequate description of the electrode's polarization behavior for SoC between 25% and 75%, if charge transfer coefficients significantly below 0.5 were used. At higher or lower SoC noticeable deviations from the experimental data were observed. Nevertheless, the Butler-Volmer equation in its presented form still seems to be a good choice for mathematical modeling of VRFBs in general.

Deviations to already reported exchange current densities in the literature might be attributed to concentration effects. The effects of vanadium and sulfate concentration on the kinetic behavior of the carbon felt or carbon paper electrode should be the focus of future work. This is especially important, since higher vanadium concentrations could reduce the current density of hydrogen evolution reaction. Unfortunately, higher vanadium concentrations lead to lower electrolyte conductivities and increase the exchange current density. Therefore, higher current densities are required to scan a sufficiently wide potential window for the evaluation of Tafel slopes. Both circumstances lead to larger potential gradients and consequently to more inhomogeneous charge transfer current densities. Thus, a compromise between high concentrations representing technically more relevant conditions and low concentrations ensuring

low potential gradients has to be made. If it reveals that the application of high vanadium concentrations is essential, the inhomogeneous charge transfer current density distribution is hardly avoidable and the evaluation of the experimental results should include parameter estimation by mathematical modeling. An alternative option might be the application of even thinner electrodes or only single carbon fibers, which would allow a homogeneous charge transfer current density distribution for high vanadium concentrations.

Acknowledgements

The authors gratefully acknowledge thyssenkrupp Industrial Solutions AG for financial support, thyssenkrupp Uhde Chlorine Engineers GmbH for technical assistance and the Energy Research Centre of Lower Saxony (EFZN) for providing the infrastructure for this project.

Appendix A. Supplementary data

Supplementary data associated with this article can be found, in the online version, at http://dx.doi.org/10.1016/j.electacta.2017.07.062.

References

[1] D.S. Aaron, Q. Liu, Z. Tang, G.M. Grim, A.B. Papandrew, A. Turhan, T.A. Zawodzinski, M.M. Mench, Dramatic performance gains in vanadium redox flow batteries through modified cell architecture, Journal of Power Sources 206 (2012) 450–453.

[2] Q.H. Liu, G.M. Grim, A.B. Papandrew, A. Turhan, T.A. Zawodzinski, M.M. Mench, High Performance Vanadium Redox Flow Batteries with Optimized Electrode Configuration and Membrane Selection, Journal of the Electrochemical Society 159 (2012) A1246–A1252.

[3] K.J. Kim, M.S. Park, Y.J. Kim, J.H. Kim, S.X. Dou, M. Skyllas-Kazacos, A technology review of electrodes and reaction mechanisms in vanadium redox flow batteries, J. Mater. Chem. A 3 (2015) 16913–16933.

[4] A.A. Shah, M.J. Watt-Smith, F.C. Walsh, A dynamic performance model for redox-flow batteries involving soluble species, Electrochimica Acta 53 (2008) 8087–8100.

[5] D. You, H. Zhang, J. Chen, A simple model for the vanadium redox battery, Electrochimica Acta 54 (2009) 6827–6836.

[6] M. Vynnycky, Analysis of a model for the operation of a vanadium redox battery, Energy 36 (2011) 2242–2256.

[7] S. Rudolph, U. Schröder, I.M. Bayanov, D. Hage, Measurement, simulation and in situ regeneration of energy efficiency in vanadium redox flow battery, Journal of Electroanalytical Chemistry 728 (2014) 72–80.

[8] A.K. Sharma, M. Vynnycky, C.Y. Ling, E. Birgersson, M. Han, The quasi-steady state of all-vanadium redox flow batteries: A scale analysis, Electrochimica Acta 147 (2014) 657–662.

[9] C. Yin, Y. Gao, S. Guo, H. Tang, A coupled three dimensional model of vanadium redox flow battery for flow field designs, Energy 74 (2014) 886–895.

[10] V.K. Yu, D. Chen, Peak power prediction of a vanadium redox flow battery, Journal of Power Sources 268 (2014) 261–268.

[11] Y. Lei, B.W. Zhang, B.F. Bai, T.S. Zhao, A transient electrochemical model incorporating the Donnan effect for all-vanadium redox flow batteries, Journal of Power Sources 299 (2015) 202–211.

[12] K.W. Knehr, E. Agar, C.R. Dennison, A.R. Kalidindi, E.C. Kumbur, A Transient Vanadium Flow Battery Model Incorporating Vanadium Crossover and Water Transport through the Membrane, Journal of The Electrochemical Society 159 (2012) A1446–A1459.

[13] E. Sum, M. Rychcik, M. Skyllas-Kazacos, Investigation of the V(V)/V(IV) system for use in the positive half-cell of a redox battery, Journal of Power Sources 16 (1985) 85–95.

[14] E. Sum, M. Skyllas-Kazacos, A study of the V(II)/V(III) redox couple for redox flow cell applications, Journal of Power Sources 15 (1985) 179–190.

[15] H. Kaneko, K. Nozaki, Y. Wada, T. Aoki, A. Negishi, M. Kamimoto, Vanadium redox reactions and carbon electrodes for vanadium redox flow battery, Electrochimica Acta 36 (1991) 1191–1196.

[16] C. Fabjan, J. Garche, B. Harrer, L. Jörissen, C. Kolbeck, F. Philippi, G. Tomazic, F. Wagner, The vanadium redox-battery: an efficient storage unit for photovoltaic systems, Electrochimica Acta 47 (2001) 825–831.

[17] G. Oriji, Y. Katayama, T. Miura, Investigations on V(IV)/V(V) and V(II)/V(III) redox reactions by various electrochemical methods, Journal of Power Sources 139 (2005) 321–324.

[18] M. Gattrell, J. Park, B. MacDougall, J. Apte, S. McCarthy, C.W. Wu, Study of the mechanism of the vanadium 4+/5+ redox reaction in acidic solutions, Journal of the Electrochemical Society 151 (2004) A123–A130.

24   *M. Becker et al. / Electrochimica Acta 252 (2017) 12–24*

[19] M. Gattrell, J. Qian, C. Stewart, P. Graham, B. MacDougall, The electrochemical reduction of VO_2^+ in acidic solution at high overpotentials, Electrochimica Acta 51 (2005) 395–407.

[20] D. Aaron, C.-N. Sun, M. Bright, A.B. Papandrew, M.M. Mench, T.A. Zawodzinski, In Situ Kinetics Studies in All-Vanadium Redox Flow Batteries, ECS Electrochemistry Letters 2 (2013) A29–A31.

[21] E. Agar, C.R. Dennison, K.W. Knehr, E.C. Kumbur, Identification of performance limiting electrode using asymmetric cell configuration in vanadium redox flow batteries, Journal of Power Sources 225 (2013) 89–94.

[22] W. Wang, X. Fan, J. Liu, C. Yan, C. Zeng, A novel mechanism for the oxidation reaction of VO2+ on a graphite electrode in acidic solutions, Journal of Power Sources 261 (2014) 212–220.

[23] W. Wang, X. Fan, J. Liu, C. Yan, C. Zeng, Unusual phenomena in the reduction process of vanadium(v) on a graphite electrode at high overpotentials, Physical Chemistry Chemical Physics 16 (2014) 19848–19851.

[24] M.-A. Goulet, M. Eikerling, E. Kjeang, Direct measurement of electrochemical reaction kinetics in flow-through porous electrodes, Electrochemistry Communications 57 (2015) 14–17.

[25] M. Becker, N. Tenhumberg, N. Bredemeyer, T. Turek, Polarization curve measurements combined with potential probe sensing for determining current density distribution in vanadium redox-flow batteries, Journal of Power Sources 307 (2016) 826–833.

[26] R. Carta, S. Palmas, A.M. Polcaro, G. Tola, Behaviour of a carbon felt flow by electrodes Part I: Mass transfer characteristics, Journal of Applied Electrochemistry 21 (1991) 793–798.

[27] S.C. Barton, Y. Sun, B. Chandra, S. White, J. Hone, Mediated Enzyme Electrodes with Combined Micro- and Nanoscale Supports, Electrochemical and Solid-State Letters 10 (2007) B96–B100.

[28] S. Umino, J. Newman, Diffusion of Sulfuric-Acid in Concentrated-Solutions, Journal of the Electrochemical Society 140 (1993) 2217–2221.

[29] A.J. Bard, G. Inzelt, F. Scholz (Eds.), Electrochemical Dictionary, Springer, Berlin, Heidelberg, 2012, pp. 913–914.

[30] S. Corcuera, M. Skyllas-Kazacos, State of Charge monitoring and electrolyte rebalancing methods for the vanadium redox flow battery, European Chemical Bulletin 1 (2012) 511–519.

[31] M. Skyllas-Kazacos, M. Kazacos, State of charge monitoring methods for vanadium redox flow battery control, Journal of Power Sources 196 (2011) 8822–8827.

[32] M. Spiro, Standard exchange current densities of redox systems at Platinum electrodes, Electrochimica Acta 9 (1964) 1531–1537.

5 Combination of impedance spectroscopy and potential probe sensing to characterize vanadium redox-flow batteries

Journal of Power Sources 446 (2020) 227349

Contents lists available at ScienceDirect

Journal of Power Sources

journal homepage: www.elsevier.com/locate/jpowsour

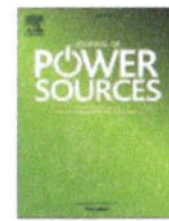

Combination of impedance spectroscopy and potential probe sensing to characterize vanadium redox-flow batteries

Maik Becker [*], Thomas Turek

Institute of Chemical and Electrochemical Engineering, Clausthal University of Technology, 38678, Clausthal-Zellerfeld, Germany

HIGHLIGHTS

- In situ potential probe sensing of liquid and solid phase.
- Validation with impedance spectroscopy model.
- Determination of NE and PE exchange current densities.
- Evaluation of bipolar plate and membrane resistances.
- Mass transport effects on overpotentials.

ARTICLE INFO

Keywords:
Solid and liquid phase potential probes
Overpotential measurement
Bipolar plate resistance
Membrane resistance
Charge-transfer resistance
Vanadium redox-flow battery

ABSTRACT

In this work, the characterization of a vanadium redox-flow battery (VRFB) under different operating conditions by means of a combination of impedance spectroscopy (EIS) and polarization curve analysis is presented. With the help of self-developed potential probes the spatially resolved overpotentials at the carbon felt electrode for the negative and positive half-cell are determined. The measured overpotentials are compared to EIS data using an analytical model describing the impedance of a porous electrode, with both methods showing very good agreement. The charge-transfer resistances of the porous electrodes vary with the SoC reaching a minimum at about 50% SoC of 0.50 $\Omega\cdot cm^2$ for the negative and 0.24 $\Omega\cdot cm^2$ for the positive half-cell. In addition to the polarization behavior of the porous electrode, solid phase potential probes provide information about the electrical resistance of the graphite plate including the contact resistance, which is determined to 0.09 $\Omega\cdot cm^2$ at a carbon felt compression rate of 9%. The electrical resistance of the Nafion® 117 membrane, which is determined at no-load conditions by evaluation of the EIS, and its resistance under charging or discharging conditions by evaluation of the liquid phase potential probe data, ranges from 0.5 to 0.9 $\Omega\cdot cm^2$.

1. Introduction

Redox-flow batteries (RFBs) have gained attention due to their independent scalability of power and capacity, resulting in potentially cost competitive applications of stationary energy storage where discharge durations of more than 4–6 h are required [1]. Among different possible electrolyte chemistries the all vanadium redox-flow battery (VRFB) is the most mature variant and systems in the megawatt power range have already been installed [2,3]. However, ongoing development in terms of cell and system performance as well as cost reduction is still necessary, to compete with other energy storage technologies [4] and to achieve the cost target of 150 $\cdot kWh^{-1}$ as given by the US Department of Energy [5].

In an effort to increase the RFB performance "zero gap" designs, with optimized pretreatment procedure and electrode thickness have been applied and led to a very high maximum power density of more than $10\,kW\cdot m^{-2}$ in small laboratory cells (5–9 cm²) [6–8]. A group at Pacific Northwest Laboratory showed a very comparable performance for a 1 kW class redox-flow battery stack with an active area of 780 cm² demonstrating that high power density can be achieved even for larger cells [9]. A recent study of Davies and Tummino indicated comparable high performance for highly compressed graphite felt electrodes [10]. Further very promising results were achieved by Zhou et al. through formation of dual scale electrodes resulting in an energy efficiency of more than 80% at current densities of $4\,kA\cdot m^{-2}$ [11]. Comparable electrodes were obtained by Liu et al. and Abbas et al. but instead of

---

* Corresponding author.
 E-mail address: maik.becker@tu-clausthal.de (M. Becker).

https://doi.org/10.1016/j.jpowsour.2019.227349
Received 30 May 2019; Received in revised form 19 September 2019; Accepted 23 October 2019
Available online 9 November 2019

M. Becker and T. Turek

Journal of Power Sources 446 (2020) 227349

etching the carbon electrode with molten potassium hydroxide more complex processes were applied [12,13].

In addition to further electrode and cell developments, it is expected that an improved characterization of an operating cell and a better understanding of limiting processes will be very helpful for performance enhancement and cost reduction. Characterization of operating cells is very often carried out by charge-discharge cycling or polarization curve analysis [14]. Some groups added potential probes to the porous electrode to receive additional information during the cell characterization. Liu et al. developed liquid phase potential probes based on thin insulated platinum wires, to measure the through plane potential distribution in the liquid phase of the positive half-cell [15]. In a subsequent paper the technique was used to validate a numerical model [16], that was focused on the concentration distribution of charged species within the electrolyte. In our previous work carbon fiber based solid phase potential probes were applied to determine potential and current density distribution for different SoC and flow rates [17]. The use of liquid phase potential probes yields the same overpotential information as ordinary reference electrode systems, but comprises two advantages. The redox potential of the electrolyte itself is used as the reference electrode system, so no Luggin capillary or additional reference electrode is required. In addition, the use of liquid phase potential probes avoids the risk of electrolyte contamination that is present for additional reference electrodes (e.g. Hg/HgSO$_4$ or Ag/AgCl). One disadvantage of potential probes can be attributed to its variable redox-potential that shifts with variation in SoC. Nevertheless, the working electrode also shows the same potential shift and the difference to the potential probe remains as the desired overpotential of the working electrode. Recording of impedance spectra and their evaluation by a suitable impedance model is another method to gain additional knowledge about the cell components. Sun et al. measured impedance spectra of small cells and could separately characterize both half-cells by using a dynamic hydrogen reference electrode. As a result they identified losses in the negative half-cell to be more significant than those in the positive half-cell [18]. In a subsequent work they measured impedance spectra of a symmetrical VRFB whose both half-cells were supplied by a single tank containing vanadium electrolyte with V^{2+} and V^{3+} species [19]. Impedance spectra of a single half-cell could be obtained by using a Hg/HgSO$_4$ reference electrode and were accessible for low frequencies without a change of state of charge (SoC). This enabled the determination of ohmic, charge-transfer and mass transport resistance. A similar setup was applied by Pezeshki et al. who investigated the effect of electrode material, flow field design and vanadium concentration [20]. Fitting the parameters of an analytical impedance model to experimental data allowed to identify exchange current density as well as ohmic resistances. Zago et al. applied a comparable setup again, in which only V^{2+} and V^{3+} species were investigated, but the evaluation was performed by using a numerical model instead of an analytical model. By adapting suitable boundary conditions, it was possible to evaluate the contribution of different mass transport mechanisms such as convection, diffusion and migration, to the total cell resistance [21].

In the present work, we demonstrate the results of impedance spectroscopy and multiple potential probe measurements of a typical flow through cell design, with VO^{2+}/VO$_2^+$ reaction on the positive and V^{2+}/V^{3+} reaction on the negative half-cell. The combination enables crosschecking of both methods and gives the opportunity to determine bipolar plate resistances, membrane resistances, and charge-transfer resistance of both half-cell reactions. A sufficiently broad range of operating parameters is ensured by variation of state of charge (SoC) and flow rate of the electrolyte. To maintain an adequate temperature range during cell characterization, the electrolyte tanks were temperature controlled and temperature probes were inserted in the inlet and outlet manifolds of the cell.

2. Experimental

2.1. Ex situ characterization of the carbon felt

Through plane resistance of the carbon felt (GFD4.6 EA, SGL Carbon GmbH, Germany) was determined in a setup of two solid 8 mm thick brass plates serving as current collectors and a piece of carbon felt placed between those. Plastic frames with different thicknesses were placed between these two plates to define a certain thickness of the compressed felt. The circular carbon felt electrode samples were cut precisely with a cutting tool to a diameter of 20 mm and single layers of the electrode were placed between the two brass plates. Solid phase probes as applied in the cell were positioned between the current collector and the carbon felt electrode on each side. The probes and the brass plates were connected to a 4-point milliohm meter (Peaktech 2705) to measure the through plane resistance of the carbon felt. By variation of plastic frame thickness different electrode compression rates could be tested. To estimate the diameter of the carbon felt fibers, microscopic images of the carbon felt were taken with a Keyence VH3-2000D microscope.

2.2. Cell design and potential probes

Experiments were conducted on the same test stand, with the same cell materials, cell design, and electrolyte (GfE GmbH, Germany, 3.9 M sulfate, 1.6 M vanadium, 0.06 M phosphate, <10 mM iron, 6 L on each side) as reported in our earlier work [17]. The cell comprised 10 × 10 cm^2 GFD4.6 EA (SGL Carbon GmbH, Germany) untreated carbon felt electrodes for negative (NE) and positive electrolyte (PE) side, PPG86 bipolar plate material (Eisenhuth Gmbh & Co KG, Germany), a Nafion® 117 membrane, soaked for at least 24 h in 1% sulfuric acid, silicone gaskets and Plexiglas® frames with a thickness leading to a carbon felt compression of 9%. A current collector of 2 mm thick brass was connected to the bipolar plate via a gold-coated nickel mesh to reduce contact resistances. 4 solid phase potential probes were assembled as reported previously [17]. Briefly, a small bunch of carbon fibers (Toolcraft, Germany) was sealed in thin lamination foil with unlaminated tips on both ends. 4 more probes were additionally covered at the internal tip both-sided with a Celgard® 3401 porous separator (25 µm thickness, 41% porosity, according to manufacturer's datasheet) as can be seen in Fig. 1 A and in the supplementary material. The porous separator prevented electrical contact to the carbon felt, but allowed the fibers of the probe to be completely wetted with electrolyte, though measuring the liquid phase potential, when contacted to a high impedance voltage measurement input. All eight potential probes were placed centrally in the inlet area of the carbon felt, approximately 15 mm above the inlet manifold (Fig. 1 C). Liquid and solid phase potential probes were arranged in four pairs, each pair comprising of a solid and a liquid phase potential probe. The first pair was positioned between NE bipolar plate and NE felt, the second was placed between NE carbon felt and membrane. The third and fourth pair were implemented on corresponding positions of PE side. NE probe signals were measured against the negative current collector electrode and PE probe against the positive current collector electrode. Direction of measurement is given by arrows in Fig. 1 D, so resulting voltages directly represent overpotentials in NE and PE electrode (e.g. measured liquid phase probe voltage at membrane side is positive for charging, as oxidation of VO^{2+} to VO$_2^+$ requires a positive overpotential). Four temperature probes (Pt100, Class A, Omega Engineering, Inc., USA) were positioned centrally in inlet and outlet manifolds of PE and NE.

2.3. Cell operation and characterization

Tests were performed on an automated test stand (FuelCon AG, Germany), that controlled flow rate, electrolyte reservoir temperature and electrical operation. Further parameters such as pressure drop across both half-cells, electrolyte conductivity, electrolyte temperature

M. Becker and T. Turek

Journal of Power Sources 446 (2020) 227349

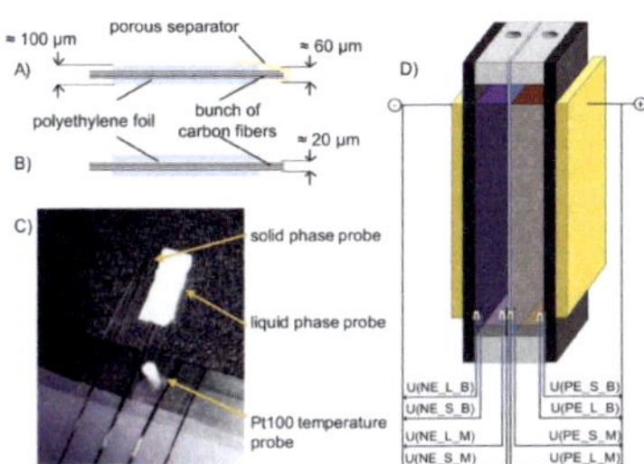

Fig. 1. Experimental setup with the solid and liquid phase potential probes. A) Liquid phase potential probes consist of carbon fibers laminated into a PE foil and covered by a Celgard 3401 separator. B) Solid phase potential probes consist of carbon fibers laminated into a PE foil. C) Picture of the potential probes with a liquid and a solid phase potential probe in front and two more below the carbon felt. The Pt100 temperature probe is placed in the inlet manifold of the cell to measure the electrolyte temperature at the inlet. D) Positions and naming of the potential probe signals shown in a cross sectional view of the cell and defined as follows:
NE/PE_#_# negative or positive electrode
##_L/S_# liquid or solid phase potential probe signal
##_##_B/M probe in the proximity of the membrane or the bipolar plate.

in hoses, fill level and potential probe signals were also recorded. Moreover, impedance spectra were measured between both current collectors in a frequency range of 0.1 Hz–100 kHz. The alternating signal was applied under galvanostatic control with a current density of $25\,\mathrm{A\cdot m^{-2}}$ or $250\,\mathrm{A\cdot m^{-2}}$, depending on the internal auto range control of the test stand and at a DC offset current density of $0\,\mathrm{A\cdot m^{-2}}$. The set point for temperature control was fixed to $25\,°\mathrm{C}$. Measurements were started with remixing of electrolyte to prevent any charge imbalance. Then electrolyte was pumped through the cell with $333\,\mathrm{ml\cdot min^{-1}}$ and charged at 1.7 V until a SoC of approximately 95% of PE was achieved. A sample of NE and PE was taken and titrated to determine SoC and total vanadium concentration [17]. Subsequently, impedance spectra were determined at six different flow rates ranging from $36\,\mathrm{ml\cdot min^{-1}}$ to $491\,\mathrm{ml\cdot min^{-1}}$, followed by data acquisition of potentiostatically controlled polarization curves at the same flow rates with alternating voltage steps for charge and discharge. After this set of polarization curves at an almost constant SoC was completed, the electrolyte was

Corcuera et al. (3.9 M total sulfate in this work to 4.2 M total sulfate in Ref. [22]). Thus, we had to adapt the temperature invariant parameters to have conformity between SoC values based on titrations and conductivity measurements. However, variations in temperature were small and its effect on SoC determination negligible. Therefore, parameters describing the linear dependence on temperature were not corrected, and the precise SoC for each polarization curve could be calculated for NE and PE, respectively.

2.4. Methodology to determine specific electrical resistance and specific surface area of the carbon felt

The specific electrical through plane resistance of the carbon felt was determined with the setup as described in the experimental section. Experimentally measured resistances R_{sample} are related to the thickness d of the compressed felt and the geometric area of the sample A_{sample} ($3.14\,\mathrm{cm^2}$) to obtain the specific electrical resistance of the felt ρ_{felt} (equation (1)). The corresponding felt compression rate CR is defined as the relative change of the felt's thickness to the original thickness of the felt d_0 (equation (2)).

$$\rho_{felt} = R_{sample} \cdot \frac{A_{sample}}{d} \tag{1}$$

$$CR = 1 - \frac{d}{d_0} \tag{2}$$

The specific surface area of the carbon felt a is required for the evaluation of the kinetic parameters of the felt. It can be calculated, if felt's porosity ε_{felt} and the carbon fiber diameter d_f are known (equation (3)):

$$a = \frac{4}{d_f}\left(1 - \varepsilon_{felt}\right) \tag{3}$$

The porosity is derived from areal weight w_{felt} and carbon fiber density ρ_{fibre} (c.f. equation (4)), that are given in the manufacturer's datasheet (GFD4.6 EA $w_{felt} = 425\,\mathrm{g\,m^{-2}}$, $\rho_{fibre} = 1.8\,\mathrm{g\,cm^{-3}}$ [23])

$$\varepsilon_{felt} = 1 - \frac{w_{felt}}{d \cdot \rho_{fibre}} \tag{4}$$

2.5. Methodology to describe impedance spectra of the porous carbon felt electrode

The impedance of the carbon felt electrode can be described by using the model of Paasch et al. [24]. We used their analytical solution for real impedance Z' and imaginary impedance Z'' of a porous electrode (c.f. equations (5) and (6)).

$$Z' = P\frac{\sqrt{\omega_1}}{\sqrt{2}}\frac{F_1\sinh(2F_1) - F_2\sin(2F_2)}{\sqrt{(k^2+\omega^2)}\left(\sinh^2(F_1)+\sin(F_2)\right)} + P\rho\sqrt{2\omega_1}\frac{F_1\sinh(F_1)\cos(F_2) - F_2\cosh(F_1)\sin(F_2)}{\sqrt{(k^2+\omega^2)}\left(\sinh^2(F_1)+\sin^2(F_2)\right)} + \frac{P\rho}{\sqrt{2\omega_1}} \tag{5}$$

discharged at 0.8 V cell voltage until the next conductivity threshold for the PE was achieved and the measurement of impedance spectra and polarization curves was started again. This method was applied for five conductivity thresholds in total, corresponding to five steps of SoC. After the completion of all measurements, SoC and total vanadium concentration were determined by titration, again. The results of titration and online measurement of conductivity during cell operation enabled an individual determination of SoC, for NE and PE during cell characterization. Linear dependence between conductivity and SoC as reported by Corcuera et al. [22] was used for the calculation of SoC. However, the composition of electrolyte in this work is not the same as applied by

$$Z'' = P\frac{\sqrt{\omega_1}}{\sqrt{2}}\frac{F_2\sinh(2F_1) + F_1\sin(2F_2)}{\sqrt{(k^2+\omega^2)}\left(\sinh^2(F_1)+\sin(F_2)\right)} + P\rho\sqrt{2\omega_1}\frac{F_2\sinh(F_1)\cos(F_2) + F_1\cosh(F_1)\sin(F_2)}{\sqrt{(k^2+\omega^2)}\left(\sinh^2(F_1)+\sin^2(F_2)\right)} \tag{6}$$

Both impedances are depending on the applied frequency ω, which is varied during the measurement. The remaining parameters P, ω_1, k and the functions F_1 and F_2, which depend on geometric, structural and

3

M. Becker and T. Turek

Journal of Power Sources 446 (2020) 227349

electrochemical properties of the porous electrode, are mainly used to simplify the equations and to improve their readability. They are explained briefly in equations (7)–(12):

$$P = \frac{1}{A(\rho + 1)} \left(\frac{\rho_{felt} + \rho_L}{2\, c_{DL}\, a} \right)^{\frac{1}{2}} \tag{7}$$

The auxiliary parameter P comprises mainly the specific resistivity of felt ρ_{felt} and liquid phase ρ_L as well as the specific double layer capacitance of the carbon felt in the electrolyte c_{DL} and its specific surface area a. A describes the geometric area of the electrode and ρ is another auxiliary parameter depending on the resistivity of carbon felt and liquid phase (equation (8)).

$$\rho = \frac{2\rho_{felt}\, \rho_L}{\rho_{felt}^2 + \rho_L^2} \tag{8}$$

The specific electrical resistance of the liquid phase ρ_L was calculated using the electrolyte conductivity σ measured during cell operation and correcting it by the Bruggemann equation (equation (9)). The porosity of the assembled felt ε_{felt} was taken from Table 1.

$$\rho_L = \frac{1}{\sigma \cdot \varepsilon_{felt}^{1.5}} \tag{9}$$

The characteristic frequency ω_1 depends on the same parameters as P, but additionally on the thickness of the compressed felt d:

$$\omega_1 = \frac{1}{d^2\, c_{DL}\, a\, (\rho_{felt} + \rho_L)} \tag{10}$$

The ratio of exchange current density to double layer capacitance is calculated with parameter k:

$$k = \frac{i_0 F}{c_{DL} R_g T} \tag{11}$$

Here, i_0 represents the exchange current density related to the carbon fiber surface, whereas F, R_g and T have their usual meaning as Faradaic constant, ideal gas constant and temperature (in the electrode).

The functions F_1 and F_2 depend on the applied frequency ω and on the already introduced parameters k and ω_1. They are defined as follows:

$$F_{1,2} \equiv \left(\frac{\sqrt{k^2 + \omega^2} \pm k}{2\omega_1} \right)^{\frac{1}{2}} \tag{12}$$

3. Results and discussion

3.1. Determination of carbon felt's resistance and specific surface area

Before evaluating electrochemical measurements, suitable values for electrical resistance, porosity and specific surface of the felt are required to perform model-based evaluation of polarization curves and impedance spectra. Results are shown in Fig. 2 A, where averaged values of a series of three individual measurements are depicted. As expected, the specific electrical resistance decreases moderately with increasing compression rate, but seems to achieve a constant or slightly increasing level with higher compression rates of more than 50%. Since the compression rate of the felt in the cell (9%) is lower than the lowest

Table 1

Parameters of the carbon felt assembled in the cell.

Parameter	Symbol	Value	Unit
Carbon fiber diameter	d_f	10	µm
Carbon felt thickness (uncompressed)	d_0	4.6	mm
Electrode thickness (compressed in cell)	d	4.21	mm
Felt porosity (compressed in cell)	ε_{felt}	94.4%	–
Specific surface area	a	$2.24 \cdot 10^4$	m^{-1}
Specific electrical through plane resistance	ρ_{felt}	2.2	$m\Omega \cdot m$

compression rate in the setup for determination of felt resistance (12%), we extrapolated the last two measured values to the compression rate of 9%. In comparison to measurements of the same carbon felt material by Schweiss et al. [25], who applied a two-point setup with gold coated contacts to measure the felt resistance for a compression rate of 20%, our results show slightly lower values. This is most likely due to the difference between four-point measurements and two-point measurements, as two-point measurements still include contact resistance in principle, even when gold coated contacts are used.

The diameter of carbon fiber was estimated to 10 µm by microscopic images of the carbon felt, that were taken with a Keyence VHS-2000D microscope as exemplarily shown in Fig. 2 B. A certain roughness of carbon fibers is visible on the microscopic images, which would lead to higher specific surface areas of the felt than calculated by equation (3). In the following evaluation of kinetic parameters we have to consider this point, especially for the comparison of reaction rate constants. The determined parameters of the carbon felt are summarized in Table 1.

3.2. Polarization curves and potential probe signals

After cell characterization was carried out as described in the experimental section, 30 data sets for polarization curves at five different levels of SoC and six different flow rates were obtained. Only selected results are shown in the main part of this work, while other polarization curves are presented in the supplementary material. During the initial charging of the electrolyte, the SoC of NE and PE increased differently, probably due to crossover effects of membrane and resulting self-discharge. The SoC of NE achieved only 80% while the PE achieved 91%. This spread between NE and PE decreased to some extent during the measurement and after the last polarization curve was recorded, the NE SoC was 6% while PE SoC reached 14%. For the sake of readability and to focus on the main aspects of cell behavior, in most figures only averaged value of SoC is indicated (unless otherwise noted). However, exact SoC of NE and PE is given in the polarization curve datasets provided in the supplementary material. Flow rate through each half-cell deviated to an extent of up to 25 ml·min⁻¹ compared to the set point in the control application of the test stand. Therefore, flow rates are shown as averaged values in figures, but as for the SoC the precise value on each side can be found in the supplementary material. In Fig. 3, polarization curve data for an averaged flow rate of 491 ml·min⁻¹ and an averaged SoC of 75% is depicted. One can see that the current density is switching between charging and discharging corresponding to the alternating cell voltage. Furthermore, a quite stable only very slightly increasing plateau is apparent for discharging process after a certain dynamic period, while a slight but steady increase of current is observed for the charging process, especially at high current densities (Fig. 3 A). This might be attributed to increasing hydrogen evolution side reactions. In addition, membrane resistance decreases for higher current densities [26]. However, this process takes a few minutes to be fully equilibrated, thus leading to a slight ongoing increase of current density. This uncertainty must be taken into account, but longer voltage steps were not applicable as they would change the SoC in the electrolyte container more considerably. Signals of PE probes and NE probes show a behavior corresponding to the current density with a quite stable only slightly increasing plateau during charge (Fig. 3 B and C), and a steep decrease during discharge reaching a more stable value at the end of a voltage step. To check if mass transport effects could strongly affect the response time of the liquid phase potential probe signal, we estimated the time required to equilibrate concentrations between the bulk of electrolyte and the electrolyte within the liquid phase potential probe by applying the mean square displacement according to equation (13).

$$x^2 = 2\, D_{eff}\, t \tag{13}$$

$$D_{eff} = D\, \varepsilon_{sep}^{1.5} \tag{14}$$

M. Becker and T. Turek

Journal of Power Sources 446 (2020) 227349

Fig. 2. A) Specific electrical resistances of the carbon felt GFD4.6 EA for different compression rates B) Microscopic image of the GFD4.6 EA taken with a Keyence VHS-2000D microscope.

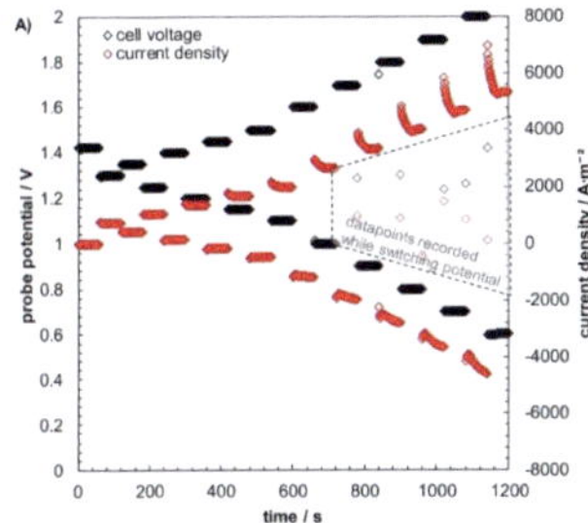

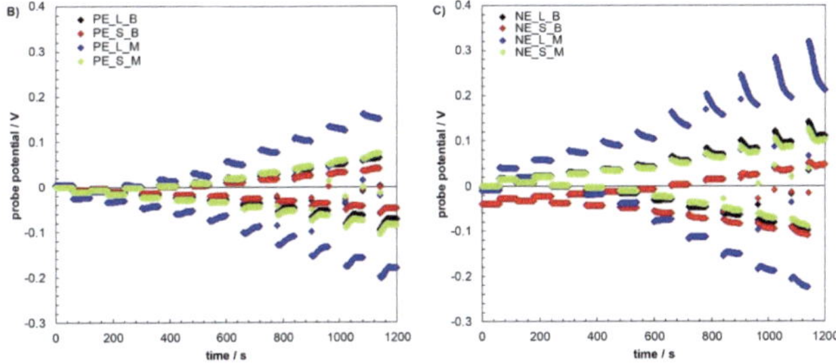

Fig. 3. A) Current density and cell voltage during the measurement of a whole polarization curve of a cell comprising of two GFD4.6 EA carbon felts at 9% compression separated by a Nafion 117 membrane. The averaged SoC (between NE and PE) is 75% at an averaged electrolyte flow rate of 491 ml·min⁻¹. B) According liquid and solid phase potential probe signals for the positive electrode. C) Liquid and solid phase potential probe signals for the negative electrode. Naming of the signals is as follows:

NE/PE_#_# negative or positive electrode
##_L/S_# liquid or solid phase potential probe signal
##_#_B/M probe in the proximity of the membrane or the bipolar plate.

Here, x represents mean displacement, D_{eff} effective diffusion coefficient in the electrolyte and t the time required for diffusion. The effective diffusion coefficient can be calculated using Bruggeman equation (14),

where D is the diffusion coefficient and ε_{sep} the separator's porosity. When we apply parameters of Table 2 (NE side) and assume that separator thickness is the required mean square displacement, an equilibration time of 5 s results. Since this is significantly lower than the 60 s

M. Becker and T. Turek

Journal of Power Sources 446 (2020) 227349

Table 2

Parameters used for the calculation of the mean square displacement.

Parameter	Symbol	Value	Unit	Reference
Diffusion coefficient (V^{2+}/V^{3+})	D	$2.4 \cdot 10^{-10}$	$m^2 \cdot s^{-1}$	[27]
Separator porosity	ε_{sep}	41%	–	Celgard's data sheet
Separator thickness	–	25	µm	Celgard's data sheet

of a constant voltage step, it is unlikely that diffusion within the potential probe considerably influences the measured signal. Since the diffusion coefficients for the VO^{2+} and VO_2^+ species are in the same order of magnitude, the time for concentration equilibration is comparable and should cause no problem on the PE side as well.

Since we are not interested in the dynamic behavior of the cell, we neglected all data except those recorded immediately before a cell voltage step, unless otherwise noted. These are considered as quasi steady state data and are used to display the polarization curves. Polarization curves obtained for three different electrolyte flow rates and all available SoCs are shown in Fig. 4 A and B. As expected, high flow rates lead to a more linear behavior of the cell voltage, because mass transport limitations become less important and ohmic losses predominate. However, it is still recognizable that, especially for discharging process, a significant change in the slope of the polarization curve occurs for different SoCs, which can be due to kinetic effects or due to changes in ohmic resistances of electrolyte or membrane. Interestingly, the slope of the polarization curve is almost constant for all SoCs during the charge process. When the flow rate is reduced, a certain curvature occurs at low SoC, corresponding to an increase of mass transport effects. At even lower flow rates one can achieve a limiting current density for charging and discharging process at high and low SoCs.

For two exemplarily polarization curves of Fig. 4 A and B at a low SoC and for high and low flow rates the corresponding potential probe signals are shown in Fig. 4 C and D. The development of liquid phase potential probe signals with current density exhibits a non-linear behavior. For all charging current densities, the dependence shows dominant kinetic effects since the current density seems to increase exponentially with the probe potential. In the case of discharge, however, two different patterns are discernible. At high flow rates (Fig. 4 D) kinetic overpotential also seems to be dominant, as the high flow rate probably attenuates any mass transport effect, but at low flow rates (Fig. 4 C) current density reaches a limiting value especially for PE, although the absolute value of potential increases. This effect indicates mass transport limitations due to low SoC and low flow rate. In contrast, solid phase potential signals for both polarization curves have a mainly linear dependence on the current density. This behavior is expected, as the potential drop in the electrically conducting carbon felt is solely caused by its ohmic resistance. Only for low flow rates and high current densities a nonlinearity occurs, which can be attributed to an inhomogeneous current density distribution [17]. It should be noted that potential probe NE_S_B, measuring solid phase potential on the bipolar plate of NE side, showed a leakage of electrolyte along the carbon fiber to the input junction of the test stand. This lead to a varying offset potential and influenced the measurement. Therefore, we used the negative value of corresponding PE potential probe as NE_S_B probe signal. This should lead to only a minor error, since bipolar plate and cell design is symmetric for both sides and the bipolar plate behaves as ohmic resistance. The measured potential probe signals can be converted to local overpotentials by subtraction of the solid phase potential from corresponding liquid phase potential. Results are shown in Fig. 4 E and F. Overpotentials at the bipolar plate are significantly lower than at the membrane side, indicating that main charge-transfer reaction is located at the membrane side for both, NE and PE reaction. The spread between membrane and bipolar plate interface overpotentials is more pronounced for the PE side than for the NE side, which indicates a more inhomogeneous overpotential distribution due to a higher exchange current density for the PE side. In contrast to that, the PE overpotential is higher than the NE overpotential when discharging at maximum current density, while it is opposite when charging. At high current densities, not only the exchange current densities but also the charge-transfer coefficients determine the overpotential. When high current densities are applied, both parameters exchange current density and charge-transfer coefficient determine the overpotential. If we consider charge-transfer coefficients as presented in our previous work [28] (c.f. Table 3), it seems that the trend of overpotentials can be explained reasonably well. For discharging the NE overpotential is lower than the PE overpotential, accordingly the NE charge-transfer coefficient is higher than the PE charge-transfer coefficient (NE: $\alpha_{an} = 0.35$, PE: $\alpha_{cat} = 0.31$). For charging the NE overpotential is higher than the PE overpotential and accordingly the charge-transfer coefficients are in reverse order (NE: $\alpha_{cat} = 0.26$, PE: $\alpha_{an} = 0.35$).

3.3. Evaluation of impedance spectra and resulting exchange current densities

For cross-check of overpotential measurement and to evaluate the kinetics in terms of exchange current densities, we measured electrochemical impedance spectra (EIS) of the full cell. Fig. 5 A and B depict the data for highest flow rate and all different SoCs in a Nyquist plot. In all cases one can recognize slightly distorted semicircles with an intersection of the real axis at high frequencies. While the intersection indicates ohmic resistance of the cell, the distorted semicircle can be explained by the combined faradaic and double-layer capacitance charging and discharging currents in the porous carbon felt electrode. A beginning of a second semicircle or a linear trend is perceptible at low frequencies, which is most probably due to mass transport effects and might be explained by a Warburg impedance. To describe the impedance behavior of the cell, we applied the model of Paasch et al. [24] as described in the methodology sections.

To obtain an appropriate representation of the measured impedance spectrum, two complex impedances Z_{NE} and Z_{PE} for NE and PE half-cell, each containing the sum of real and imaginary part according to equations (5) and (6), are connected in series with an ohmic cell resistance $R_{mem+bpp}$ and an inductance L_{dis}, as shown in the insertion of Fig. 5 C. The parameters L_{dis}, $R_{mem+bpp}$, i_{0NE}, i_{0PE} and c_{DL} were adapted to match the complex impedance of the model to the measured impedance by a least squares method for frequencies between 11 kHz and 50 Hz. Specific capacitance of carbon felt electrode was assumed to be identical for NE and PE side, since the same electrode is used in similar electrolytes. Deviations in specific capacitance arising from different redox potentials of NE and PE were neglected. The information from potential probes, that PE reaction is more facile than NE reaction, was used in the way that the model referred higher exchange current density always to PE side and the lower to NE side. In Fig. 5 C an exemplary impedance spectrum and its corresponding fit is shown, revealing a good agreement. The adapted exchange current densities for different flow rates but constant SoC were used to build arithmetic average and standard deviation, which are shown as dots and vertical error bars in Fig. 5 D. As expected [29,30], the exchange current density is highest for a SoC of approximately 50% and decreases for higher and lower SoC. In general, the dependence of exchange current density on concentration of two oxidized and reduced species, can be described in the following way [31–34] (equation (15)):

$$i_0 = Fkc_{ref} \left(\frac{c_{ox}}{c_{ref}} \right)^{\nu_{ox}} \left(\frac{c_{red}}{c_{ref}} \right)^{\nu_{red}} = i_{00} \left(\frac{c_{ox}}{c_{ref}} \right)^{\nu_{ox}} \left(\frac{c_{red}}{c_{ref}} \right)^{\nu_{red}} \tag{15}$$

i_{00} corresponds to the standard exchange current density, which is the exchange current density at standard concentrations c_{ref} for both oxidized and reduced species, c_{red} and c_{ox}, respectively. It is proportional

M. Becker and T. Turek

Journal of Power Sources 446 (2020) 227349

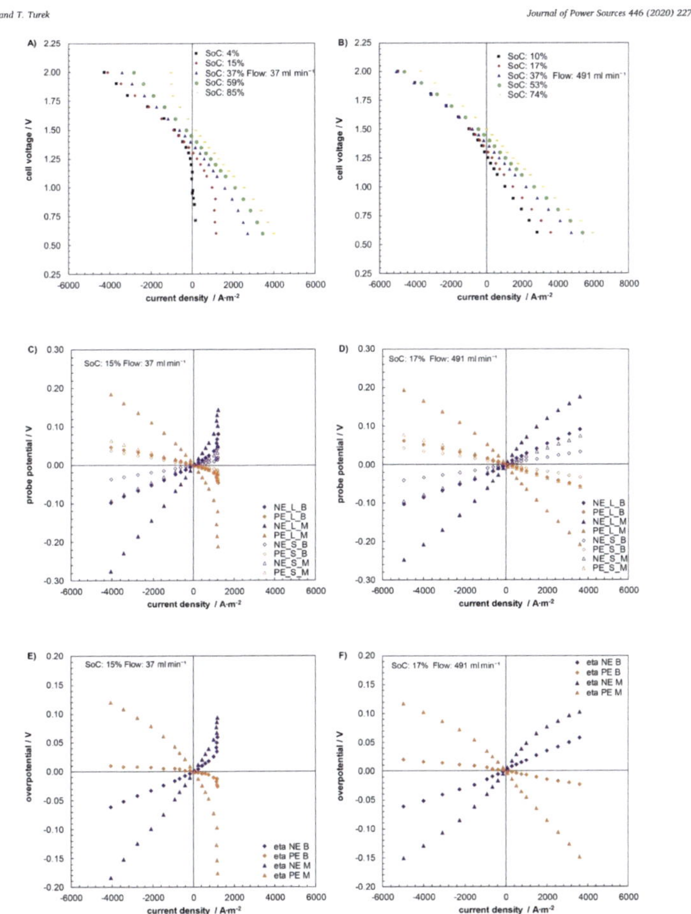

Fig. 4. A) and B) Polarization curves at different flow rate and SoC obtained with untreated GFD4.6 EA carbon felt at 9% compression on NE and PE half-cell separated by a Nafion® 117 membrane. C) Probe signals for solid and liquid phase potential probes for a SoC of 16% and a flow rate of 37 ml·min⁻¹. D) Probe signals for solid and liquid phase potential probes for a SoC of 17% and a flow rate of 491 ml·min⁻¹. E) and F) Resulting overpotentials calculated as the difference from solid phase and liquid phase potential probe signals under the same conditions as shown in C) and D), respectively. Shown SoC and flow rates are averaged values over NE and PE. Naming of the signals is as follows:

NE/PE_#_# negative or positive electrode
##_L/S_# liquid or solid phase potential probe signal
##_#_B/M probe in the proximity of the membrane or the bipolar plate
eta_#_# overpotential, difference between solid and liquid phase potential probe.

M. Becker and T. Turek

Journal of Power Sources 446 (2020) 227349

Table 3

Charge-transfer coefficients for NE and PE reaction based on the data in Ref. [28] and averaged for the two different materials investigated.

	NE	PE
α_{an}	0.35	0.35
α_{cat}	0.26	0.31

to the reaction rate constant and can be determined from it, or vice versa. The reaction orders ν_{red} and ν_{ox} are depending on the charge-transfer coefficients α_{an} and α_{cat} as well as on reaction orders of each species for anodic and cathodic reactions, $\nu_{ox,an}$, $\nu_{ox,cat}$ and $\nu_{red,an}$, $\nu_{red,cat}$, respectively as shown in equations (16) and (17) [32]. Often these dependencies are simplified under the first assumption that $\nu_{ox,an}$ and $\nu_{red,cat}$ are zero (e.g. VO^{2+} is not taking part in the cathodic reaction of the PE, and VO_2^+ is not taking part in the anodic reaction of the PE), which seems to be reasonable. In an additional second assumption the reaction orders $\nu_{ox,cat}$ and $\nu_{red,an}$ are set to one (e.g. the VO^{2+} concentration is of first order in the cathodic reaction of the PE and the VO_2^+ concentration is of first order in the anodic reaction of the PE), thus reaction order ν_{ox} equals α_{cat} and ν_{ox} equals α_{an}.

$$\nu_{ox} = \alpha_{cat}\nu_{ox,an} + \alpha_{an}\nu_{ox,cat} \tag{16}$$

$$\nu_{red} = \alpha_{cat}\nu_{red,an} + \alpha_{an}\nu_{red,cat} \tag{17}$$

Unfortunately, only a few publications on the experimental investigation of reaction orders are available. Wang et al. showed that the reaction order of VO^{2+} for the oxidation reaction on pyrolytic graphite and glassy carbon has a value of two [35]. In a following paper these authors investigated the reaction order of VO_2^- reduction on a platinum electrode and also found an order of two [36]. This indicates, that the often applied second assumption is not generally true for the reaction in PE half-cell. However, we expect that the first assumption is valid without having further evidence from literature. Although the described experimental setup and corresponding results are not ideally designed for determination of reaction orders, it is still possible to check the effect of different reaction orders on exchange current density and to identify more probable values when comparing to experimental data. Therefore, we applied reaction orders of one and of two for the reacting species (VO^{2+} for oxidation in PE, VO_2^+ for reduction in PE, V^{2+} for oxidation in NE and V^{3+} for reduction in NE). The required charge-transfer coefficients were taken from our previous work [28] (c.f. Table 3).

Standard exchange current densities were adapted to fit to measured data and the resulting dependencies on SoC are shown as lines in Fig. 5 D. For exchange current densities of PE half-cell, it seems that a reaction

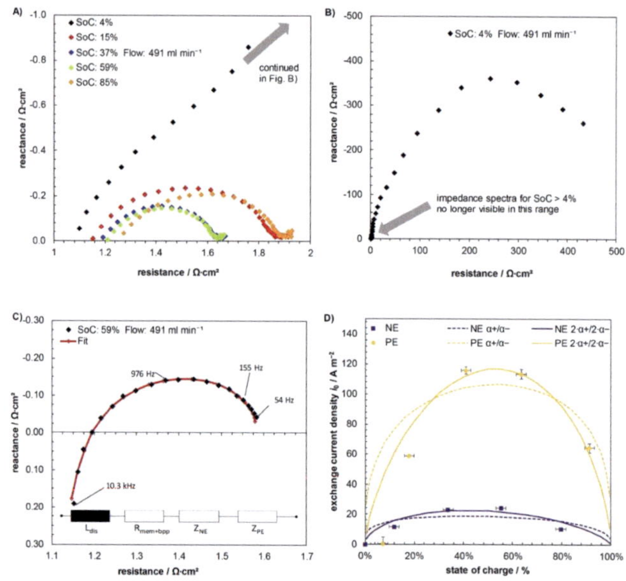

Fig. 5. A) Nyquist plot for different SoC at the highest average flow rate of 491 ml·min⁻¹. The galvanostatic impedance spectrum was measured between positive and negative current collector at OCV conditions (DC offset current density = 0 A·m⁻²). B) Nyquist plot for low SoC of 4%, shown over the entire range of complex resistance. C) Comparison of the measured (59% average SoC and average flow rate of 491 ml·min⁻¹) Nyquist plot to the modeled Nyquist plot according to the equivalent circuit model shown in the insertion. D) Measured exchange current densities for the NE and PE reaction in dependence on the SoC (dots) and interpolated curves according to equation (15) (lines).

8

M. Becker and T. Turek

Journal of Power Sources 446 (2020) 227349

Table 4

Reaction rate constants from literature compared to this work (CF = carbon felt, CP = carbon paper. *calculated with reaction orders of one, ** calculated with reaction orders of two for reacting species).

Material type	Material	Supplier	NE k_0/m·s^{-1}	PE k_0/m·s^{-1}	Reference
CP	10AA	SGL Carbon	$3.09 \cdot 10^{-7}$	$1.36 \cdot 10^{-5}$	[18] recalculated by Ref. [37]
CP	HCP030 N	Shanghau Hensen	$1.13 \cdot 10^{-5}$	$1.04 \cdot 10^{-5}$	[38]
CP	10AA	SGL Carbon	$7.6-15 \cdot 10^{-7}$	n.a.	[20]
CF	GFA5	SGL Carbon	$1.54 \cdot 10^{-7}$	$1.40 \cdot 10^{-7}$	[39] recalculated by Ref. [40]
CF	n.a.	Alfa Aesar	$7.40 \cdot 10^{-8}$	$1.10 \cdot 10^{-7}$	[41] recalculated by Ref. [40]
CF	GFD4.6	SGL Carbon	$1.00 \cdot 10^{-8}$	$6.20 \cdot 10^{-7}$	[25]
CF	GFD3	SGL Carbon	$2.6-3.2 \cdot 10^{-7}$	n.a.	[20]
CF	GFD4.6	SGL Carbon	$2.2 \cdot 10^{-7}$	$1.3 \cdot 10^{-6}$	this work*
CF	GFD4.6	SGL Carbon	$3.0 \cdot 10^{-7}$	$1.6 \cdot 10^{-6}$	this work**

order of two results in a better description than for a reaction order of one, especially for a SoC around 50%. A comparable picture evolves for NE half-cell, since a reaction order of two again shows a better agreement with experimental data than a reaction order of one. Although no clear conclusion can be drawn from our results, they indicate that the assumption of a first-order reaction might be inappropriate and that more detailed work on the kinetic characterization of carbon felt electrodes is required to clarify the reaction mechanisms. Nevertheless, resulting reaction rate constants for both assumptions are shown in Table 4 and the difference between the values for a first-order and a second-order reaction is less than 50% for NE and PE reactions. If compared to other reaction rate constants for carbon felt and carbon paper materials from literature, this deviation is quite small, since the range for reaction rate constants varies over three orders of magnitude. A good comparison can be made with data from Schweiss et al. [25], since the same electrode material was used. Reaction rate constants of these authors are more than two times lower for PE and more than 20 times lower for NE reaction. We think that this deviation is due to various reasons. First, characterization methods cyclic voltammetry and impedance spectroscopy do not necessarily yield the same result. Second, the methods to determine specific surface area were different and BET surface area measurements by Schweiss et al. resulted in an almost two times higher surface area compared to our calculation of the specific surface area assuming perfectly round carbon fibers, most probably because the BET method accounts for surface roughness. Third, our evaluation method employing the model of Paasch et al. [24] considers an inhomogeneous current distribution, which can be essential for the NE electrode with its lower electrolyte conductivity compared to PE. To indicate the importance of inhomogeneous current distribution we can calculate the reaction rate constants with and without considering current distribution. If we assume an exchange current density i_0 of $30 \, \text{A} \cdot \text{m}^{-2}$ for NE reaction at 50% SoC and other parameters remain as described for our cell, the model-free evaluation would result in a reaction rate constant of just 21% of the result considering inhomogeneous current distribution. This clearly shows that EIS data evaluation without support of a model can lead to an underestimation of the reaction rate constants. A good agreement can be found for reaction rate constant of Pezeshki et al. [20], who used a similar model and investigated the same carbon felt material, but in a thinner configuration of only 3 mm thickness and only for NE reaction.

3.4. Comparison of EIS results with potential probe signals

An evaluation of the whole potential probe data would require at least a simple model considering potential and current density distributions to eventually determine reaction rate constants that can be compared to those from EIS data. This is out of scope of this work. Nevertheless, both measurements can be compared in a simpler way, when potential probe data is limited to the low overpotential range and a linear relationship between current and overpotential is used to calculate the exchange current densities that can be compared to the EIS data. The model of Paasch et al. [24] is applied again. The real part of complex impedance Z' as described by equation (5) turns into an apparent electrode polarization resistance $R_{ct,app}$, if the applied frequency is set to zero (pure DC behavior). This resistance considers current density and potential distribution in the porous electrode and comprises kinetic and ohmic voltage losses. It can be converted into an apparent exchange current density $j_{0,app}$ according to equation (18).

$$j_{0,app} = \frac{R_g T}{F R_{ct,app}} \tag{18}$$

The apparent exchange current density $j_{0,app}$ is a 0D parameter that describes utilization of the electrode and is related to geometrical area of the electrode. It should not be confused with exchange current density i_0 that is related to carbon fiber surface. In case of an ideal porous electrode without inhomogeneous current density distribution apparent exchange current density would depend only on exchange current density i_0 as well as on the product of electrode thickness d and specific surface area of the electrode a. This would give an ideal exchange current density $j_{0,ideal}$ (c.f. equation (19)), which is always greater than or equal $j_{0,app}$. The product ad could be defined as area enhancement factor, because it describes the ratio of total electrode area to geometric surface area which in our case of GFD4.6 EA carbon felt is around $95 \, \text{m}^2 \, \text{m}^{-2}$.

$$j_{0,ideal} = i_0 \cdot ad \tag{19}$$

Since EIS data were obtained close to open circuit potential of the cell and in a frequency range of more than 50 Hz, it is very likely that vanadium concentration on the carbon fiber surface equals the concentration in electrolyte bulk and that the concentration gradient between inlet and outlet of the cell is negligible. Therefore, a low difference between inlet and outlet SoC during polarization curve measurement must be maintained by appropriate operational conditions. Furthermore, a good mass transport from bulk phase to carbon fiber surface must be ensured as well. Such requirements can only be fulfilled at high flow rates of approximately $500 \, \text{ml} \cdot \text{min}^{-1}$ [17]. In addition, high flow rate reduces temperature variations from cell inlet to outlet to less than 1 K compared to temperature variations of more than 3 K at a low flow rate of approximately $50 \, \text{ml} \cdot \text{min}^{-1}$. This eliminates temperature influences on evaluated data of polarization curve measurements and ensures a good comparability between EIS and potential probe data.

Another determination of the apparent exchange current density $j_{0,app}$ can be derived from the slope of overpotentials of the electrode. The appropriate overpotential must be selected to achieve comparable results between EIS and potential probe data. The EIS model considers electrode overpotential from the interface of bipolar plate and felt electrode (solid phase) to the interface between membrane and carbon felt electrode (liquid phase). This overpotential corresponds to the difference of liquid phase potential at carbon felt interface to membrane $U_{L,M}$ to solid phase potential of carbon felt at the interface to the bipolar plate $U_{S,B}$ for NE and PE half-cell respectively (c.f. equation (20) and Fig. 6 A). We define it as the apparent felt overpotential $\eta_{felt,app}$.

$$\eta_{felt,app} = U_{L,M} - U_{S,B} \tag{20}$$

M. Becker and T. Turek

Journal of Power Sources 446 (2020) 227349

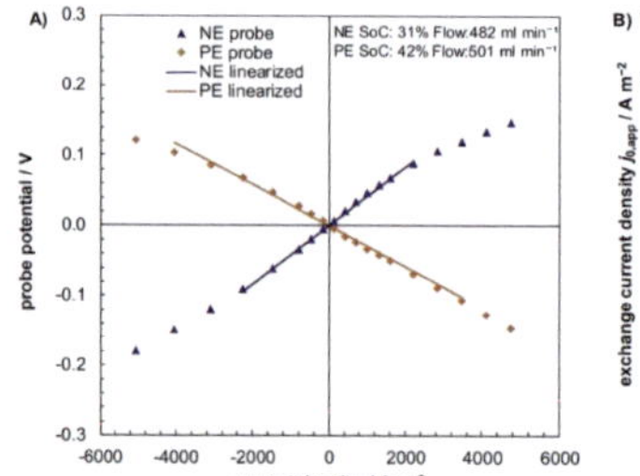
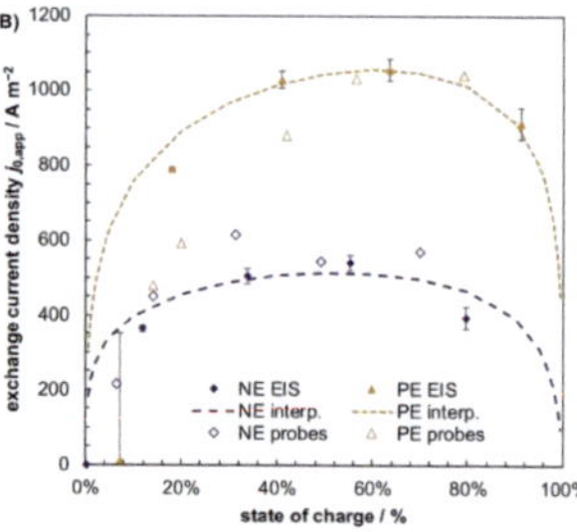

Fig. 6. A) Measured felt overpotentials and linearized overpotential current density behavior for NE and PE exemplarily shown for a mild SoC level and high flow rates as shown in the figure. B) Apparent exchange current densities for NE and PE. EIS data points are derived from the impedance spectra data evaluation, "interp," were calculated based on the interpolation of exchange current densities (c.f. Fig. 5 D), "probes" data points are derived from the felt linearized overpotential current density dependence.

As can be seen exemplary in Fig. 6 A, felt overpotential below an absolute value of approx. 0.1 V shows a good agreement to a linear fit in this range. The slope of linear fit is the apparent polarization resistance as already determined by EIS. Correspondingly, this can be converted again into apparent exchange current density by using equation (18). Additionally to apparent exchange current densities from EIS and potential probe data, we calculated them by interpolating the exchange current densities i_0 (c.f. equation (15) and Fig. 5 D) by assuming a reaction order of two, since it provided a more satisfactory agreement between measurement and interpolation as (c.f. previous section). The comparison of apparent exchange current densities for different SoC is shown in Fig. 6 B. There is a good agreement for NE data. However, PE data show a certain deviation for SoC lower 50% between interpolated data and overpotential data. The EIS data point for the lowest SoC of PE is quite low and has a high standard deviation itself, but it fits better to overpotential data in the low SoC region. Therefore, it may be the case that interpolation of exchange current densities of PE reaction still needs further adaption to describe kinetics more accurately.

3.5. Quantification of bipolar plate, membrane and electrode resistances

Potential probe sensing and impedance spectroscopy allow the analysis of the contribution of different electrode components to overall cell resistance. Certainly, such an evaluation of potential probe signals should be carried out at high flow rates to ensure a homogeneous current density distribution. The solid phase potential probe in contact to the carbon felt and to the bipolar plate allows the determination of the bipolar plate resistance including its contact resistance. The linear curvature of the polarization curve of this solid phase potential probe in the PE half-cell indicates a homogeneous current density distribution for all SoCs, since all curves are similar in a narrow range (Fig. 7 A). For the reported cell setup one yields a bipolar plate area specific resistance of $0.09\,\Omega\cdot cm^2$, which is slightly higher than the electrical resistance calculated by the manufacturers data, which is $0.04\,\Omega\cdot cm^2$. The difference is most likely caused by the additional contact resistance at the bipolar plate carbon felt interface.

The membrane resistance can be determined by using two liquid phase potential probe signals in proximity to the membrane in NE and PE half-cell. As their signals include all losses due to bipolar plate resistance, electrode resistances and kinetic overpotentials but not the membrane resistance, one can obtain the membrane voltage drop by subtracting liquid phase potential probe signals from total voltage losses of the cell. These total voltage losses are simply the difference of the cell voltage under application of a current $U_{cell}(j)$ compared to open circuit voltage of the cell U_{OCV}. The membrane voltage drop ΔU_{mem} is therefore calculated as follows (equation (21))

$$\Delta U_{mem}(j) = U_{OCV} - U_{cell}(j) - U_{PE,L,M}(j) + U_{NE,L,M}(j) \tag{21}$$

where $U_{NE,L,M}(j)$ and $U_{PE,L,M}(j)$ are the signals of liquid phase probes for NE and PE half-cell at the applied current density j. The resulting membrane voltage drops for different SoC are shown in Fig. 7 B. As a guidance for the eye we added linear trends for charging and discharging currents, where the slope of the trend is the membrane resistance. It becomes apparent, that charge and discharge resistances are different and membrane resistance during charging is lower than the resistance during discharging. Additionally, evaluation of EIS data allows a determination of membrane resistance under open circuit conditions. The ohmic resistance $R_{mem + bpp}$, which is shown in Fig. 5 C is the sum of the membrane resistance and the resistance of both bipolar plates. Since the potential probe evaluation yields bipolar plate resistance, it can be used to calculate membrane resistance. The resulting membrane resistances under open circuit conditions during charging and discharging are shown in Fig. 7 C. In general, membrane resistance during charging is always lower than membrane resistance during discharging and under open circuit conditions. Furthermore, open circuit and charging resistance show a decreasing trend for lower SoC. The discharging resistance behaves somewhat different, since it is quite constant, besides one value at 20% SoC, which is significantly higher than the rest. This behavior is in agreement with our prior work on characterization of ion exchange membranes [26]. However, the value of the Nafion© 117 membrane resistance during discharging presented in this work is higher than the value from Ref. [26], which might be attributed to different measurement technique and higher current density range as well as to equilibration times in this work (60 s of a constant voltage) that might be too short to achieve steady-state membrane properties.

As already discussed, electrode polarization comprises of carbon felt and electrolyte resistances as well as on kinetics, therefore the apparent charge-transfer resistance of the electrode $R_{ct,app}$ for NE and PE reaction (c.f. equation (18)) can be used to quantify electrode contribution to total cell resistance. The comparison of bipolar plate resistance (calculated for two bipolar plates), membrane resistance and electrode resistances is shown in Fig. 7 D for a SoC of approximately 50%. The sum of all single resistances, averaged for charging, discharging and open circuit conditions, yields a total cell resistance of $1.63\,\Omega\cdot cm^2$ which is in

M. Becker and T. Turek

Journal of Power Sources 446 (2020) 227349

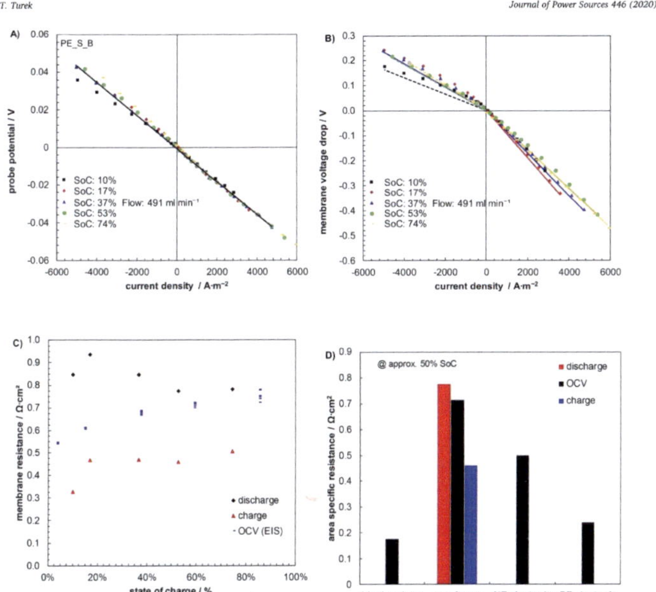

Fig. 7. Polarization curves and resistances for high flow rate conditions of approx. 500 ml·min^{-1} for the reported cell setup (GFD4.6 EA untreated carbon felt electrode at 9% compression on both sides, Nafion® 117 membrane, PPG86 bipolar plate). A) Polarization curve of the solid phase potential probe signal placed at the bipolar plate carbon felt interface of the PE half-cell. B) Membrane voltage drop for different SoC as calculated by equation 21. C) Membrane resistances from EIS evaluation compared to membrane resistances for charging and discharging as obtained from the slopes of the membrane voltage drops. D) Comparison of the polarization resistances of the different cell components for a SoC of approx. 50%. The bipolar plate resistance is the sum of both bipolar plates applied in the single cell.

good agreement with total cell resistance of 1.64 Ω·cm^2, that can be obtained from the slope of the polarization curve as shown in Fig. 4 B for a SoC of 53% and for absolute current densities below 1000 A m^{-2} (to represent open circuit voltage conditions). For the reported cell setup, the membrane resistance is the dominant loss factor, but resistance of NE electrode is quite significant as well. PE electrode losses and losses at the bipolar plate, however, seem to be of minor importance. An optimization of cell performance can therefore be carried out by using a thinner, lower resistance membrane or porous separator, as well as thinner and more intensively compressed carbon felt electrodes, which would result in a more even current density distribution and lower contact resistances.

3.6. Mass transport effects on overpotential

The flow rate has a significant effect on the behavior of a redox flow battery as it determines the electrolyte utilization in combination with the current density, influences the mass transport effects in the carbon felt electrode and leads to a pressure drop across the cell. The dependence of the electrode overpotential measured at the felt membrane interface on different flow rates during discharging is exemplarily

shown in Fig. 8 A and C, for NE and PE half-cell respectively. The SoC is low and for all polarization curves quite constant. In an overpotential window below 60 mV or for current densities lower than approximately 1000 A·m^{-2} all curves overlap in a narrow range. For higher overpotentials current densities increase the more intensively the cell is flowed through. Basically, this can be explained by larger concentration gradients for higher current densities and thus lower concentrations of the reacting species on the carbon fiber surface, which reduce the exchange current density and increase the overpotential. However, mass transport effects that lead to a concentration gradient must be divided into two effects. The first is convective mass transport along the flow through direction of the cell. If this effect is limited, concentration of reacting species at the outlet can become quite low, resulting in a low local current density in this area and an increased local current density in the inlet area (since the integral of all local current densities must remain constant). The second effect is mass transport from bulk phase of the electrode to carbon fiber surface by diffusion through the diffusion boundary layer. If this effect is dominant, current densities for a fixed overpotential would increase for higher flow rates due to a reduced diffusion boundary layer but would not depend significantly on the position within the cell.

M. Becker and T. Turek

Journal of Power Sources 446 (2020) 227349

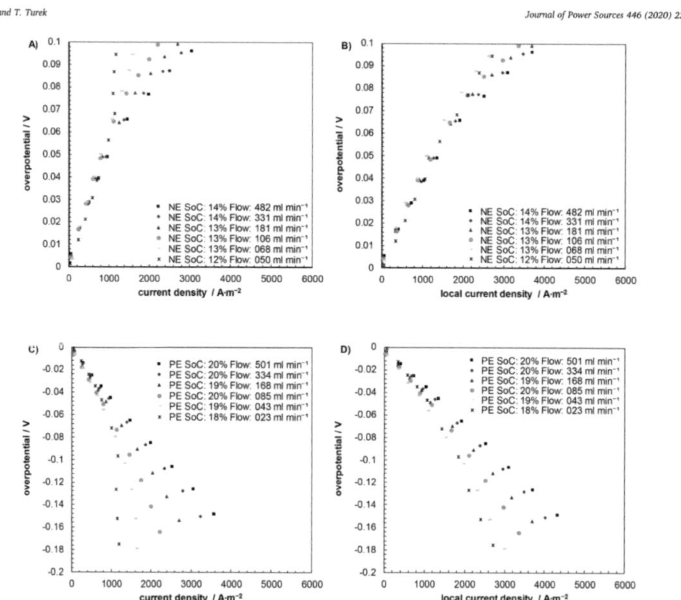

Fig. 8. Overpotentials at the carbon felt membrane interface for low SoC and variation of flow rate for NE and PE. A) NE overpotential versus the current density of the cell. B) NE overpotential versus the local current density at the probe's position. C) PE overpotential versus the current density of the cell. D) PE overpotential versus the local current density at the probe's position.

To distinguish whether the convective or diffusive mass transport effect is limited, the overpotential signals can be used and combined with the method of dynamic resistances as published in our earlier work [17] to determine the local current density at the position of the potential probes. Briefly, dynamic behavior during alternating voltage steps is used to determine a dynamic resistance of respective solid phase probe, and dynamic resistances of all polarization curves are averaged. Dynamic resistance is used as shunt resistor to calculate local current density by using corresponding solid phase potential signal under steady-state conditions. The local steady-state current densities for each solid phase probe (PE_S_B, PE_S_M and NE_S_M) are then averaged to yield the averaged local current density. Since all probes are essentially at the same position of the flow through direction of the cell, it can be assumed that this local current density is similar for the liquid phase probes. The resulting dependence of overpotential and local current density is shown in Fig. 8 B and D, for NE and PE respectively. Transition from current density to local current density generally increases the current density range, which is due to a partially limited convective mass transport. The difference between high and low flow rates decreases, which is particularly significant for NE half-cell. Since inlet SoC is quite constant for all curves and potential probes are placed in the inlet area of the cell, the SoC at position of the potential probes is quite comparable. Therefore, the dependence of the overpotential on the local current density indicates a sufficient diffusive mass transport for NE,

while the PE side seems to be more restricted by diffusive mass transport. This can at least partly be explained by the approximately 70% higher diffusion coefficient of V^{2-} compared to VO_2^+ [42] and is very probably also influenced by different electrolyte viscosities, but a more quantitative investigation goes beyond the scope of this work.

4. Conclusions

In this work, we could demonstrate the application of carbon fiber based potential probes for liquid and solid phase in a VRFB. By positioning the probes at different interfaces in the cell, we could distinguish between overpotentials for both half-cells and for positions at felt membrane and felt bipolar plate interfaces. The overpotential in the proximity of the bipolar plate was substantially lower, hence the electrochemical reaction is slow in this area. A significantly higher overpotential at felt membrane interface indicates this zone to be the main reaction zone. Additionally, kinetic data of NE and PE reaction were gathered by the evaluation of impedance spectroscopic data and validated by the comparison to the potential probe data. The NE reaction rate constant is in the range of 2.2–$3.0 \cdot 10^{-7}$ m·s⁻¹ and PE reaction rate constant is in the range of 1.3–$1.6 \cdot 10^{-6}$ m·s⁻¹ indicating a considerably faster reaction on PE side. Evaluation of impedance spectroscopic data of carbon felt electrodes should be carried out using a model considering porous electrode structure, to account for effects of uneven current

M. Becker and T. Turek

Journal of Power Sources 446 (2020) 227349

density distribution. A further analysis was carried out to determine membrane resistances for charging, discharging and open circuit conditions, whereas the membrane resistance during charging is the lowest. This behavior fits well to the results obtained in our previous work using other experimental methods. The evaluated resistances of different cell components were compared and allow the determination of performance limiting components, i.e. membrane and NE electrode. The variation of flow rate indicates a situation where higher flow rates are beneficial for higher current densities. We could determine the local current density and therefore we could compensate the effects of convective mass transport limitation. Thus, it becomes clear that both diffusive and convective mass transport limitations are present for low flow rates, but diffusive mass transport limitation is of minor importance for the NE half-cell. Quantification of kinetic properties and resistances of cell components provides a good starting point for a numerical redox-flow battery model, that will be presented in a forthcoming paper and with which the quantification of mass transports effects occurring at lower flow rates becomes possible.

Acknowledgement

The authors gratefully acknowledge thyssenkrupp Industrial Solutions for financial support and the Research Center for Energy Storage Technologies of the Clausthal University of Technology for providing the infrastructure.

Appendix A. Supplementary data

Supplementary data to this article can be found online at https://doi.org/10.1016/j.jpowsour.2019.227349.

References

[1] M. Skyllas-Kazacos, M.H. Chakrabarti, S.A. Hajimolana, F.S. Mjalli, M. Saleem, Progress in flow battery Research and development, J. Electrochem. Soc. 158 (2011) R55–R79, https://doi.org/10.1149/1.3599565.

[2] C. Minke, T. Turek, Materials, system designs and modelling approaches in techno-economic assessment of all-vanadium redox flow batteries – a review, J. Power Sources 376 (2018) 66–81, https://doi.org/10.1016/j.jpowsour.2017.11.058.

[3] K. Yano, S. Hayashi, T. Kumamoto, T. Shibata, K. Yamanishi, K. Fujikawa, Development and demonstration of redox flow battery system, SEI Tech. Rev. (2017) 22–28.

[4] M.L. Perry, A.Z. Weber, Advanced redox-flow batteries: a perspective, J. Electrochem. Soc. 163 (2016) A5064–A5067, https://doi.org/10.1149/2.0101601jes.

[5] U.S. Department of Energy, Grid energy storage. https://www.energy.gov/sites/prod/files/2014/09/f18/Grid%20Energy%20Storage%20December%202013.pdf, 2013. (Accessed 24 June 2018).

[6] D.S. Aaron, Q. Liu, Z. Tang, G.M. Grim, A.B. Papandrew, A. Turhan, T. A. Zawodzinski, M.M. Mench, Dramatic performance gains in vanadium redox flow batteries through modified cell architecture, J. Power Sources 206 (2012) 450–453, https://doi.org/10.1016/j.jpowsour.2011.12.026.

[7] Q.H. Liu, G.M. Grim, A.B. Papandrew, A. Turhan, T.A. Zawodzinski, M.M. Mench, High performance vanadium redox flow batteries with optimized electrode configuration and membrane selection, J. Electrochem. Soc. 159 (2012) A1246–A1252, https://doi.org/10.1149/2.051208jes.

[8] J. Houser, A. Pezeshki, J.T. Clement, D. Aaron, M.M. Mench, Architecture for improved mass transport and system performance in redox flow batteries, J. Power Sources 351 (2017) 96–105, https://doi.org/10.1016/j.jpowsour.2017.03.083.

[9] D. Reed, E. Thomsen, B. Li, W. Wang, Z. Nie, B. Koeppel, V. Sprenkle, Performance of a low cost interdigitated flow design on a 1 kW class all vanadium mixed acid redox flow battery, J. Power Sources 306 (2016) 24–31, https://doi.org/10.1016/j.jpowsour.2015.11.089.

[10] T.J. Davies, J.J. Tummino, High-performance vanadium redox flow batteries with graphite felt electrodes, C 4 (2018) 8, https://doi.org/10.3390/c4010008.

[11] X.L. Zhou, Y.K. Zeng, X.B. Zhu, L. Wei, T.S. Zhao, A high-performance dual-scale porous electrode for vanadium redox flow batteries, J. Power Sources 325 (2016) 329–336, https://doi.org/10.1016/j.jpowsour.2016.06.048 .

[12] Y. Liu, Y. Shen, L. Yu, L. Liu, F. Liang, X. Qiu, J. Xi, Holey-engineered electrodes for advanced vanadium flow batteries, Nano Energy 43 (2018) 55–62, https://doi.org/10.1016/j.nanoen.2017.11.012.

[13] S. Abbas, H. Lee, J. Hwang, A. Mehmood, H.-J. Shin, S. Mehboob, J.-Y. Lee, Y. Ha, A novel approach for forming carbon nanorods on the surface of carbon felt electrode by catalytic etching for high-performance vanadium redox flow battery, Carbon 128 (2018) 31–37, https://doi.org/10.1016/j.carbon.2017.11.066.

[14] Y.A. Gandomi, D.S. Aaron, J.R. Houser, M.C. Daugherty, J.T. Clement, A. M. Pezeshki, T.Y. Ertugrul, D.P. Moseley, M.M. Mench, Critical review—experimental diagnostics and material characterization techniques used on redox flow batteries, J. Electrochem. Soc. 165 (2018) A970–A1010, https://doi.org/10.1149/2.0601805jes.

[15] Q. Liu, A. Turhan, T.A. Zawodzinski, M.M. Mench, In situ potential distribution measurement in an all-vanadium flow battery, Chem. Commun. 49 (2013) 6292–6294, https://doi.org/10.1039/C3CC42092B.

[16] Y.A. Gandomi, D.S. Aaron, T.A. Zawodzinski, M.M. Mench, In situ potential distribution measurement and validated model for all-vanadium redox flow battery, J. Electrochem. Soc. 163 (2016) A5188–A5201, https://doi.org/10.1149/2.0211601jes.

[17] M. Becker, N. Bredemeyer, N. Tenhumberg, T. Turek, Polarization curve measurements combined with potential probe sensing for determining current density distribution in vanadium redox-flow batteries, J. Power Sources 307 (2016) 826–833, https://doi.org/10.1016/j.jpowsour.2016.01.011.

[18] D. Aaron, C.-N. Sun, M. Bright, A.B. Papandrew, M.M. Mench, T.A. Zawodzinski, In situ kinetics studies in all-vanadium redox flow batteries, ECS Electrochem. Lett. 2 (2013) A29–A31, https://doi.org/10.1149/2.001303eel.

[19] C.-N. Sun, F.M. Delnick, D.S. Aaron, A.B. Papandrew, M.M. Mench, T. A. Zawodzinski, Resolving losses at the negative electrode in all-vanadium redox flow batteries using electrochemical impedance spectroscopy, J. Electrochem. Soc. 161 (2014) A981–A988, https://doi.org/10.1149/2.045406jes.

[20] A.M. Pezeshki, R.L. Sacci, F.M. Delnick, D.S. Aaron, M.M. Mench, Elucidating effects of cell architecture, electrode material, and solution composition on overpotentials in redox flow batteries, Electrochim. Acta 229 (2017) 261–270, https://doi.org/10.1016/j.electacta.2017.01.056.

[21] M. Zago, A. Casalegno, Physically-based impedance modeling of the negative electrode in All-Vanadium Redox Flow Batteries: insight into mass transport issues, Electrochim. Acta 248 (2017) 505–517, https://doi.org/10.1016/j.electacta.2017.07.166.

[22] S. Corcuera, M. Skyllas-Kazacos, State-of-Charge monitoring and electrolyte rebalancing methods for the vanadium redox flow battery, Eur. Chem. Bull. 1 (2012) 511–519, https://doi.org/10.17628/ecb.2012.1.511-519.

[23] SGL Carbon GmbH, SIGRACET® and SIGRACELL® - Components for Flow Batteries, 2012.

[24] G. Paasch, K. Micka, P. Gersdorf, Theory of the electrochemical impedance of macrohomogeneous porous electrodes, Electrochim. Acta 38 (1993) 2653–2662, https://doi.org/10.1016/0013-4686(93)85083-B.

[25] R. Schweiss, Influence of bulk fibre properties of PAN-based carbon felts on their performance in vanadium redox flow batteries, J. Power Sources 278 (2015) 308–313, https://doi.org/10.1016/j.jpowsour.2014.12.081.

[26] K. Schafner, M. Becker, T. Turek, Membrane resistance of different separator materials in a vanadium redox flow battery, J. Membr. Sci. 586 (2019) 106–114, https://doi.org/10.1016/j.memsci.2019.05.054.

[27] T. Yamamura, N. Watanabe, T. Yano, Y. Shiokawa, Electron-transfer kinetics of Np^{3+}/Np^{4+}, NpO_2^+/NpO_2^{2+}, V^{2+}/V^{3+}, and VO^{2+}/VO_2^+ at carbon electrodes, J. Electrochem. Soc. 152 (2005) A830, https://doi.org/10.1149/1.1870794.

[28] M. Becker, N. Bredemeyer, N. Tenhumberg, T. Turek, Kinetic studies at carbon felt electrodes for vanadium redox-flow batteries under controlled transfer current density conditions, Electrochim. Acta 252 (2017) 12–24, https://doi.org/10.1016/j.electacta.2017.07.062.

[29] E. Hollmann, Factors Influencing Kinetics of Electrode Processes in Vanadium Redox Flow Batteries, University of Tennessee Honors Thesis Projects, 2015. https://trace.tennessee.edu/utk_chanhonoproj/1818/.

[30] S. Wagner, A. Oberland, T. Turek, Analytical approach for evaluation of lithium-ion battery cells, Energy Technol. 4 (2016) 1543–1549, https://doi.org/10.1002/ente.201600137.

[31] S. Schuldiner, M. Rosen, The exchange current density vs. Concentration relation and its use in a rigorous determination of solution purity, J. Electroanal. Chem. Interfacial Electrochem. 35 (1972) 1–6, https://doi.org/10.1016/S0022-0728(72)80287-0.

[32] R. Parsons, The transfer coefficient in electrode reactions, Croat. Chem. Acta 42 (1970) 281–291.

[33] A.J. Bard, G. Inzelt, F. Scholz (Eds.), Electrochemical Dictionary, second ed., Springer-Verlag, Berlin Heidelberg, 2012.

[34] R. Parsons, Electrode reaction orders, transfer coefficients and rate constants. Amplification of definitions and recommendations for publication of parameters, Pure Appl. Chem. 52 (1980) 233–240, https://doi.org/10.1351/pac198052010233.

[35] W. Wang, X. Fan, J. Liu, C. Yan, C. Zeng, A novel mechanism for the oxidation reaction of VO2+ on a graphite electrode in acidic solutions, J. Power Sources 261 (2014) 212–220, https://doi.org/10.1016/j.jpowsour.2014.03.053.

[36] W. Wang, X. Fan, Y. Qin, J. Liu, C. Yan, C. Zeng, The reduction reaction kinetics of vanadium(V) in acidic solutions on a platinum electrode with unusual difference compared to carbon electrodes, Electrochim. Acta 283 (2018) 1313–1322, https://doi.org/10.1016/j.electacta.2018.07.071.

[37] A. Bourke, M.A. Miller, R.P. Lynch, X. Gao, J. Landon, J.S. Wainright, R.F. Savinell, D.N. Buckley, Electrode kinetics of vanadium flow batteries: contrasting responses of V^{II}-V^{III} and V^{IV}-V^{V} to electrochemical pretreatment of carbon, J. Electrochem. Soc. 163 (2016) A5097–A5105, https://doi.org/10.1149/2.0131601jes.

[38] X.W. Wu, T. Yamamura, S. Ohta, Q.X. Zhang, F.C. Lv, C.M. Liu, K. Shirasaki, I. Satoh, T. Shikama, D. Lu, S.Q. Liu, Acceleration of the redox kinetics of VO_2^+/VO^{2+} and V^{3+}/V^{2+} couples on carbon paper, J. Appl. Electrochem. 41 (2011) 1183, https://doi.org/10.1007/s10800-011-0343-7.

13

M. Becker and T. Turek

Journal of Power Sources 446 (2020) 227349

[39] E. Agar, C.R. Dennison, K.W. Knehr, E.C. Kumbur, Identification of performance limiting electrode using asymmetric cell configuration in vanadium redox flow batteries, J. Power Sources 225 (2013) 89–94, https://doi.org/10.1016/j.jpowsour.2012.10.016.

[40] J. Friedl, U. Stimming, Determining electron transfer kinetics at porous electrodes, Electrochim. Acta (2017), https://doi.org/10.1016/j.electacta.2017.01.010.

[41] M.Z. Jelyani, S. Rashid-Nadimi, S. Asghari, Treated carbon felt as electrode material in vanadium redox flow batteries: a study of the use of carbon nanotubes as electrocatalyst, J. Solid State Electrochem. 21 (2017) 69–79, https://doi.org/10.1007/s10008-016-3336-y.

[42] Z. Jiang, K. Klyukin, V. Alexandrov, Structure, hydrolysis, and diffusion of aqueous vanadium ions from Car-Parrinello molecular dynamics, J. Chem. Phys. 145 (2016) 114303, https://doi.org/10.1063/1.4962748.

6 Mass transport parameter estimation and model validation for vanadium redox-flow batteries

6.1 Introduction

In the previous chapter it was shown how the application of potential probes can be utilized to determine membrane and bipolar plate resistances in a redox-flow battery. Additionally, the potential probes enabled a validation method to check the exchange current densities from impedance spectroscopy by the overpotentials measured with the probes, which showed an acceptable agreement between both methods. Eventually, a first evaluation of mass transport effects could be carried out, indicating a mass transport limitation by convection and diffusion for both electrodes. In this chapter the mass transport effects will be determined in more detail by mathematical modelling of the whole cell and parameter estimation of an appropriate mass transport correlation. The determined charge transfer coefficients presented in chapter 4 are used in the model to describe the overpotential current density behavior.

Mass transport is a frequently occurring process in the field of chemical engineering, where a variety of approaches can be applied to describe the underlying effects quantitively. Mass transport in a carbon felt electrode of a redox-flow battery takes place at the interface area between the liquid electrolyte phase and the solid phase of the carbon felt's fibers. A common assumption to describe the process is the separation in a bulk electrolyte phase and a diffusion boundary layer as illustrated in Fig. 6-1. The typical design of a redox-flow battery forces the electrolyte to flow through the carbon felt electrode, leading to a region between two neighboring fibers where the flow velocity is high, defined as bulk phase of the electrolyte. At the surface of the carbon fiber a no-slip condition limits the electrolyte flow velocity to zero and a thin layer of non-moving electrolyte phase is assumed to persist with a constant thickness around each carbon fiber, defined as diffusion boundary layer. Since no convection can provide mass transfer to the carbon fiber surface during application of faradaic currents, diffusion is expected to be the only effect, that can transport reacting vanadium species to the carbon fiber surface. The mass flux J_i to the carbon fiber surface can be described by Fickian diffusion as shown in equation (6-1),

$$J_i = \frac{c_{\text{bulk},i} - c_{\text{surf},i}}{\delta_i} \cdot D_i \qquad (6\text{-}1)$$

where $c_{\text{bulk},i}$ represents the concentration of an exemplary species i in the bulk phase and $c_{\text{surf},i}$ describes the concentration at the surface of the carbon felt fiber. δ_i is the thickness of the diffusion boundary layer and D_i defines the diffusion coefficient of species i in the surrounding electrolyte. In general, the thickness of the diffusion boundary layer decreases with higher flow rates, since they lead to higher shear-rates at the fiber electrolyte interface, resulting in higher mass transfer rates to the carbon fiber surface. Another possibility to describe the mass transport is the use of the mass transfer coefficient $k_{\text{m},i}$, simply representing the ratio of diffusion coefficient and thickness of the diffusion boundary layer (c.f. equation 6-2):

$$k_{\text{m},i} = \frac{D_i}{\delta_i} \qquad (6\text{-}2)$$

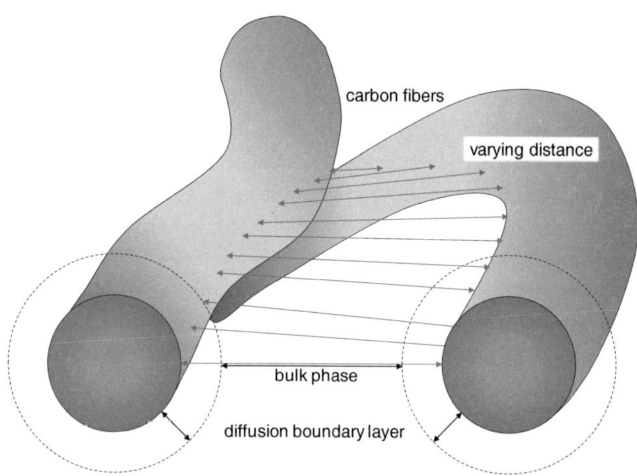

Figure 6-1: Schematic drawing of the diffusion layer between two exemplary sketched fibers.

On the microscopic scale a mathematical description of the diffusion boundary layer in a carbon felt electrode with its randomly ordered carbon fibers would be quite complex, since the distance between two fibers is constantly changing, the alignment of the fibers can be parallel or perpendicular to the flow or anything in between, some of the fibers may have contact to each other leading to partially blocked areas and the diffusion boundary layer itself probably won't have a constant thickness around a single fiber. Thus, an analytical approach to quantify the diffusion boundary layer seems to be a quite challenging task. However, a very common way to describe the dependence between flow rate and mass transport is the use of a dimensionless averaged value for the mass transport coefficient of a certain geometry depending on a dimensionless flow rate [125]. Sherwood Sh_i and Reynolds Re numbers, describing the dimensionless mass transport and the flow velocity are defined as shown in equations 6-3 and 6-4.

$$Sh_i = \frac{k_{\mathrm{m},i} \cdot L}{D_i} \tag{6-3}$$

$$Re = \frac{u \cdot \rho_{\mathrm{el}} \cdot L}{\mu_{\mathrm{el}}} \tag{6-4}$$

L is a characteristic length of the applied geometry, u is the superficial flow velocity through the geometry ρ_{el} and μ_{el} are the density and the dynamic viscosity of the electrolyte, respectively. The dependence between Reynolds number and Sherwood number can be expressed in a general form as depicted in equation 6-5.

$$Sh_i = a_{\mathrm{Sh}} Re^{b_{\mathrm{Sh}}} Sc_i^{c_{\mathrm{Sh}}} \tag{6-5}$$

In this equation a_{Sh}, b_{Sh} and c_{Sh} are parameters that can be derived analytically for more simple geometries or experimentally for complex structures or geometries. Usually the exponents b_{Sh} and c_{Sh} are in the range of ⅓ to ½ but higher values are also possible [126,127]. Sc_i is the Schmidt number representing the ratio of kinematic viscosity and diffusion coefficient as defined in equation 6-6 for a given exemplary species i:

$$Sc_i = \frac{\mu_{el}}{\rho_{el} \cdot D_i} \qquad (6\text{-}6)$$

Several publications describe the investigations about the mass transport to carbon fiber electrodes or single carbon fibers. One of the first works was published by Kinoshita et al., who investigated the reduction reaction of bromine in a zinc halide electrolyte that was pumped through the electrode [128]. The electrodes under investigation were two carbon felt electrodes with a porosity of 86% and 90% and quite thick carbon fibers with a diameter of 25.4 µm. In their work the carbon fiber diameter was chosen as the characteristic length for the calculation of Reynolds and Sherwood number, as it was repeated in most of the following publications (c.f. table 6-1). Schmal et al. determined the mass transport to single carbon fibers in a 1 M potassium hydroxide solution containing small amounts of dissolved potassium hexacyanoferrate(III), that underwent reduction at the carbon fiber [94]. They applied the carbon fiber diameter as characteristic length as well and their mass transport correlation is applied in a large number of mathematical models describing vanadium redox-flow batteries [49,70,73,81,86,87,93,99,101,102,109,113,114,121,129–138]. Other mass transport correlations are summarized in table 6-1, whereas all correlations apply the carbon fiber diameter as the characteristic length. A different approach can be found in the two correlations of table 6-2, where the hydraulic diameter d_h of the carbon felt represents the characteristic length. It can be derived from the carbon fiber diameter d_f and the porosity ε of the electrode, according to equation 6-7 [139]:

$$d_h = d_f \frac{\varepsilon}{1-\varepsilon} \qquad (6\text{-}7)$$

The hydraulic diameter or sometimes called pore diameter is the mean distance between all carbon fibers. If the fibers would be arranged perfectly parallel and evenly distributed, this would imply a maximum diffusion length of the half of the hydraulic diameter. Following this assumption the mass transport correlation of Xu et al. comprises a constant offset of two for the Sherwood number [140]. Independent from the reliability of this assumption, it is important to consider that carbon fibers in the carbon felt electrode are structured quite randomly with some regions where two or more fibers are in contact and other regions where the distance to the adjacent fiber is considerably higher than the mean distance.

Table 6-1: Mass transport correlations for carbon based fibrous materials – fiber diameter applied as characteristic length. [*]Schmidt number was calculated based on the given diffusion coefficient and on a kinematic viscosity estimated form the graphical information on flow velocity and Reynolds numbers. [**]Parameters fitted to the published graphical data points. [†]Schmidt number derived by the ratio of Peclet to Reynolds number from published data sets (Peclet number: $Pe_i = d_f \cdot u / D_i$ → $Pe_i = Re \cdot Sc_i$).

Reference	Correlation(s)	Electrolyte	Electrode	Schmidt number
Kinoshita et al. [128]	$Sh = 1.29\, Re^{0.72}$ $Sh = 1.01\, Re^{0.61}$ $0.01 < Re < 0.4$	6 to 62 mM Br_2 in 2 M $ZnBr_2$ + 1M $ZnCl_2$ + 3 M KCl	Carbon felt, type CH, Fiber Materials Inc. Fiber diameter: 25.4 µm	~900[*]
Schmal et al. [94]	$Sh = 7\, Re^{0.4}$ $0.04 \leq Re \leq 0.2$	0.1 to 10 mM $K_3Fe(CN)_6$ in 1 M KOH	Carbon fiber, Le Carbone Lorraine AGT/ FT10 000 Fiber diameter: 8 µm	~1600 (estimated to be similar to Vatistas et al.)
Vatistas et al. [141]	$Sh = 4.26\, Re^{0.64}$ [**] $0.003 < Re < 0.19$	1 or 3 mM $K_3Fe(CN)_6$ in 3 or 9 mM $K_4Fe(CN)_6$ + 0.5 M KOH	Carbon felt, GFD 5, SIGRI GmbH Fiber diameter: 8 µm	1393– 2264[†]
You et al. [142]	$Sh = 1.68\, Re^{0.9}$ $0.08 < Re < 1.43$	50 mM $FeCl_2$ in 0.75 M $FeCl_3$ + 2 M HCl	Carbon felt, type n.a. Fiber diameter: 10 µm	1522
Barton et al. [143] (flow through FT and interdigitated ID)	$Sh = 0.004\, Re^{0.75} Sc^{0.51}$ $0.005 < Re < 0.4$ (FT) $Sh = 0.018\, Re^{0.68} Sc^{0.50}$ $0.0006 < Re < 0.05$ (ID)	0.5 M $FeCl_2$ + 0.5 M $FeCl_3$ in 2 M HCl	Carbon paper, Sigracet 29AA, SGL Carbon Fiber diameter: 7 µm	3800– 24000
Kok et al. [144]	$Sh = 0.9\, Re^{0.4} Sc^{0.4}$ (rounded) $10^{-6} < Re < 1$	n.a.	Carbon felt, GFD, SGL Carbon Fiber diameter: 8.2 µm	n.a.

Table 6-2: Mass transport correlations for carbon based fibrous materials – hydraulic diameter applied as characteristic length. The subscript $_{dh}$ indicates the Sherwood and Reynolds numbers to be related to the hydraulic diameter.

Reference	Correlation	Electrolyte	Electrode	Schmidt number
Carta et al. [139]	$Sh_{dh} = 3.19\,Re^{0.69}$ $1 < Re_{dh} < 15$	2 mM $K_3Fe(CN)_6$ in 100 mM $K_4Fe(CN)_6$ + 1 M KNO_3 and 1 mM $CuSO_4$ in 0.1 M H_2SO_4	Carbon felt, SIGRI GmbH Fiber diameter: 11 µm	1203 and 1365
Xu et al. [140]	$Sh_{dh} =$ $2 + 1.534\,Re_{dh}^{0.912}$ $0.3 < Re_{dh} < 2.4$	0.125 M to 1.0 M V @90% SoC in 3M H_2SO_4	GFA, SGL Carbon GmbH Fiber diameter: n.a.	n.a.

For a comprehensive comparison all mass transport correlations are presented in Figure 6-1 A). The correlations of Barton et al. and Kok et al. were calculated based on an exemplary Schmidt number of 5000. The correlations of Carta et al. and Xu et al. had to be adapted to fit into the same diagram, because the characteristic lengths are different from the other correlations. Therefore, the ratio of the fiber diameter to the hydraulic diameter is used to adapt the equations, which itself is just a function of the porosity that was assumed to be 94%. The subscript $_{dh}$ denominates Sherwood and Reynolds numbers that are related to the hydraulic diameter of the felt. Reynolds and Sherwood numbers without subscripts represent the ones related to the fiber diameter, unless otherwise noted.

$$Re_{dh,i} = Re\,\frac{d_h}{d_f} = Re\,\frac{\varepsilon}{1-\varepsilon} \tag{6-8}$$

$$Sh = Sh_{dh}\,\frac{d_f}{d_h} = Sh_{dh}\,\frac{1-\varepsilon}{\varepsilon} \tag{6-9}$$

In Fig. 6-2 it is apparent, that the range of Sherwood numbers can vary around three orders of magnitude, depending on the choice of mass transport correlations and the range of Reynolds numbers. In order to correct the effect of different electrolyte compositions and thus different Schmidt numbers within the listed publications, it is necessary to add a Schmidt number dependence to each correlation and subsequently compare it for a constant Schmidt number. Only two correlations comprise the parameter c_{Sh} (exponent of the Schmidt number), which varies between 0.4 and 0.5, thus it would seem reasonable to assume a mean value of 0.45 for c_{Sh}. Consequently, the parameter a_{Sh} needs to be corrected as well, to fit to a Schmidt number dependent mass transport correlation according to the following equation:

$$a_{Sh} = \frac{a_{Sh'}}{Sc'^{0.45}} \tag{6-10}$$

The values a_{Sh}' and Sc' indicate the parameters reported in the publication, whereas a_{Sh} represents the corrected mass transport correlation parameter. The resulting correlations are compared in figure 6-2 B) for an exemplary Schmidt number of 20972, which corresponds to a completely discharged positive electrolyte containing only VO^{2+} species. The figure still indicates a broad spread of Sherwood numbers with a very slight concentration of correlations in the range of Barton et al. (interdigitated ID), Kinoshita et al. and You et al. [128,142,143]. Some of the presented correlations might be less reliable due to a variety of reasons. Kok et al. [144] based the correlations on a Lattice Boltzmann method for the computation of fluid-flow within the 3D structure of a carbon felt derived by tomographic methods. Since the evaluation is purely based on a numerical approach and comprises a high deviation to other experimental correlations its reliability is expected to be rather low. The correlation Barton et al. presented for flow through cell (FT) gives quite low Sherwood numbers [143]. Their cell setup consisted of two small (14x16 mm²) layers of SGL Sigracet® GDL29AA carbon paper material, with a compressed thickness of approximately 0.3 mm. The active cell area defined by the gasket had dimensions of 15x17 mm², so a 1 mm bypass channel was likely to exist along the edge of the carbon paper. Since the pressure drop across a pure flow through cell setup is larger than the pressure drop across an interdigitated flow field design (ID), the bypassing effect is presumably significantly larger for the FT than for the ID design. This can explain the difference between both mass transport correlations quite well and shows that the FT correlation is not that reliable. The interdigitated correlation is presumably more reliable, however the bypassing may have had an effect on these experiments as well. Thus, a true Sherwood number higher than predicted by the correlation of Barton et al. is likely. Nevertheless, the evaluation of viscosity changes and its resulting effect on Schmidt number dependence is unlikely to be influenced by the bypassing, since the laminar flow regime within the cell maintained a constant ration of flow through the electrode and through the bypassing channel. The correlation of Xu et al. [140] starts with the principally good assumption that the minimum Sherwood number is limited to two, thus the maximum diffusion boundary layer is limited to the half of the average distance between two carbon fibers. Unfortunately, the experiments of Xu et al. were carried out on a cell setup with parallel flow fields. This setup leads to low flow velocity through the electrode, since the permeability of the carbon paper is usually much lower than the permeability of a flow channel, thus underestimating the true mass transport in the carbon felt. Although very often applied in mathematical models describing redox-flow batteries, the correlation of Schmal et al. [94] bases on measurements using single fiber electrodes. The extrapolation of their mass transport correlation to carbon felt electrodes with their inhomogeneous structure is at least questionable. The fibers in carbon felt electrodes might block surface areas of adjacent fibers, so a lower Sherwood number than predicted by Schmal et al. is probable. The correlations of Kinoshita et al., Vatistas et al. and You et al. [128,141,142] measured in flow through cell setups remain for the range of most probable mass transfer correlations in carbon felt electrodes (marked as grey box in Fig. 6-2 B)). When considering the Reynolds numbers that are experimentally accessible in this work, the correlation of You et al. must be neglected as well. The exponents

b_{Sh} for the Reynolds number of the remaining correlations vary between 0.61 and 0.72 and build an average value of 0.66 that is used for further evaluation.

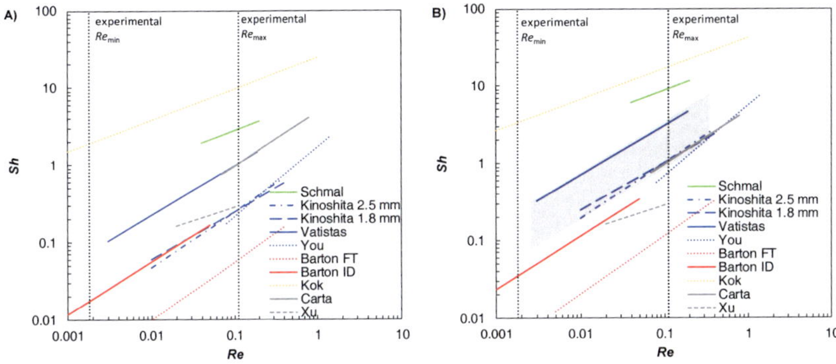

Figure 6-2: A) Mass transport correlations for carbon electrodes as reported in different publications of Schmal et al. [94], Kinoshita et al. (two electrodes) [128], Vatistas et al. [141], You et al. [142], Barton et al. with flow through cell design (FT) and interdigitated cell design (ID) [143], Kok et al. [144], Carta et al. [139], Xu et al. [140], whereas the Schmidt number was fixed to 5000 and the porosity of the electrode was assumed to be 94%. The Sherwood and Reynolds number are related to the fiber diameter as characteristic length. B) Appropriate mass transport correlations as in A), but including a Schmidt number dependence for all correlations with a Schmidt number of 20972 (completely discharged positive electrolyte). They grey box indicates the most probable range for correct mass transport correlation as discussed in the text.

6.2 Setup of the mathematical model

The model in this is based on the model of Shah et al. published in 2008 [41] and Ma et al. from 2011 [113], with some adaptions and some simplifications where appropriate. The model itself is designed as a two-dimensional model describing the cell processes of a single cell of a redox-flow battery. The model dimensions are placed in the cross section of a flow-through cell design as presented in Fig. 6-3, since this allows the description of the potential and concentration distribution in the through plane (x) and in the direction of electrolyte flow (y). The third dimension is neglected, because a sufficiently good design of the electrolyte manifolds guarantees a homogeneous flow distribution and therefore the behavior in each layer of the third dimension is identical. The model focusses on the processes within the electrodes and neglects effects of membrane crossover, therefore describing the membrane solely as an ohmic resistance. The electrodes are described in terms of mass transport by convection, diffusion and reaction and in terms of potential and current density distribution by modelling the solid and liquid phase resistances connected by nonlinear resistances that depend on concentration and overpotential. Mass transport effects due to migration of vanadium are neglected, since the mobility of protons in the electrolyte is significantly faster compared to the vanadium species. The negative electrolyte (NE) electrode is denominated by the through plane dimension x, whereas the positive electrolyte (PE) electrode is denominated by the through plane dimension x′.

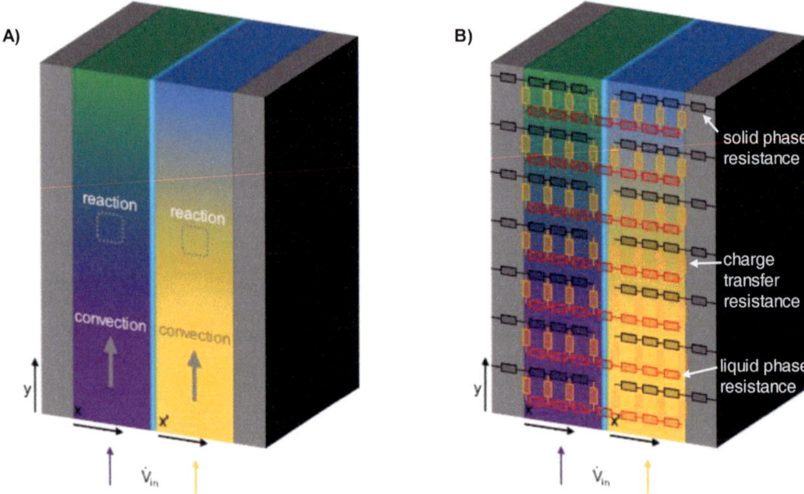

Figure 6-3: A) Model setup of the redox-flow battery for mass transport and B) for the potential and current density distribution. Bipolar plate and membrane are considered as pure ohmic resistances while the carbon felt in the NE and PE region is a combination of electronic and ionic resistance connected by the electrochemical reaction as a nonlinear resistance, that is depending on mass transport properties.

The different equations required for the model can be separated in several groups. For sake of readability the equations are presented in tables, one for each group and only the key variables are mentioned in the text. Parameters and variables not mentioned in the text are listed in table 6-11 and table 6-12. The basic reactions in the model are the redox reaction between V^{2+} and V^{3+} in the NE and the redox reaction between VO^{2+} and VO_2^+ in the PE. Any effects by water or protons are neglected and so the reaction is simplified as shown in equation 6-11 and 6-12, whereas the vanadium species in the PE are denominated in the simpler version of V^{4+} and V^{5+}. The oxidation reaction is defined to result in a positive current. The indices, stoichiometric coefficients v and reaction orders (v_{an}, v_{cat}) used in several equations are listed in table 6-3, whereas the reaction orders of two were derived from the results of the chapter 5.

$$\text{NE:} \qquad V^{2+} \leftrightarrows V^{3+} + e^- \qquad E_{ref,NE} = -0.255 \text{ V vs. SHE} \qquad (6\text{-}11)$$
$$\text{PE:} \qquad V^{4+} \leftrightarrows V^{5+} + e^- \qquad E_{ref,PE} = 1.004 \text{ V vs. SHE} \qquad (6\text{-}12)$$

Table 6-3: Indices and reaction orders applied in the model

Index / Order	Formula	
Reaction	$r = [\text{NE}, \text{PE}]$	(6-13)
Vanadium species	$i = [\text{V}^{2+}, \text{V}^{3+}, V^{4+}, V^{5+}] \equiv [1, 2, 3, 4]$	(6-14)
Stoichiometric coefficient	$\upsilon = \begin{bmatrix} -1 & 1 & 0 & 0 \\ 0 & 0 & -1 & 1 \end{bmatrix}$	(6-15)
Anodic reaction order	$\upsilon_{\text{an}} = \begin{bmatrix} 2 & 0 & 0 & 0 \\ 0 & 0 & 2 & 0 \end{bmatrix}$	(6-16)
Cathodic reaction order	$\upsilon_{\text{cat}} = \begin{bmatrix} 0 & 2 & 0 & 0 \\ 0 & 0 & 0 & 2 \end{bmatrix}$	(6-17)

The potential gradient in x-dimension within the solid phase and the liquid phase of the electrode is depending on its specific electrical resistance and the local current density in the corresponding phase. The specific electrical resistance of the liquid phase is determined by the Bruggeman equation based on the conductivity of the electrolyte as seen in table 6-4.

The calculation of current densities is summarized in table 6-5. The gradient of the current densities in both phases is depending on the specific surface area and on the transfer current density, which is the current density that is achieved on the surface of a carbon fiber in a differential volume element. It can be calculated by the Butler-Volmer equation as already shown in chapter 4. The resulting charge transfer coefficients of chapter 4 are averaged and applied for the Butler-Volmer equation, as it was already done in chapter 5. The overpotential is derived by the difference between solid phase and liquid phase potential minus the equilibrium potential, calculated with the surface concentration of the contributing vanadium species. Required exchange current densities are calculated based on the surface concentrations accordingly to chapter 5 (or equation 6-26).

Mass transport processes are summarized in table 6-6. The change in concentration is calculated by the sum of a convective flux in y-direction and a source term due to the transfer current density. Diffusion in x- or y-direction is assumed to be negligible due to quite high flow velocities and low diffusion coefficients of the vanadium species. The mass transport from the bulk of the electrolyte to the carbon fiber surface is described by the Sherwood correlation as already described in the first section of this chapter. By appropriate settings in the solver the dynamic setup of equation 6-27 can be deactivated (time derivatives are set to zero), resulting the cell to be operated in a steady-state mode, which is sufficient for the description of the recorded polarization curves.

Some auxiliary equations are required to describe the parameters of the carbon felt and the electrolyte. They are summarized in table 6-7. The specific surface area of the electrolyte is derived by the carbon fiber diameter and the porosity [139], whereas the porosity can be calculated by the areal weight and the bulk fiber density of the electrode. The compression ratio is defined as the change from initial electrode thickness to the thickness under compression. The state of charge (SoC) is the ratio of charged species concentration to the total vanadium concentration. Viscosity and conductivity of the electrolyte are assumed to be linear functions of SoC (c.f. [43]) and solely depend on the inlet concentration of the cell neglecting changes of SoC within the cell. This assumption is obviously not entirely correct, but the typical cell operation leads only to minor changes in SoC across the cell and the effect of SoC changes on conductivity and viscosity is mitigated by a certain offset value for both properties independent from SoC, hence this assumption causes only a small error.

Table 6-4: Formulas applied in the model for the description of the potential distribution

Type of formula	Formula	
Potential gradient in solid phase	$\dfrac{\mathrm{d}\varphi_{S,r}}{\mathrm{d}x} = -\rho_S \cdot j_{S,r}$	(6-18)
Potential gradient in liquid phase	$\dfrac{\mathrm{d}\varphi_{L,r}}{\mathrm{d}x} = -\rho_{L,r} \cdot j_{L,r}$	(6-19)
Specific electrical resistance of electrolyte	$\rho_{L,r} = \dfrac{1}{\kappa_{el,r} \cdot \varepsilon^{1.5}}$	(6-20)

Table 6-5: Formulas applied in the model for the description of the current density distribution

Type of formula	Formula	
Gradient of current density in solid phase	$\dfrac{\mathrm{d}j_{S,r}}{\mathrm{d}x} = a \cdot i_r$	(6-21)
Gradient of current density in liquid phase	$\dfrac{\mathrm{d}j_{L,r}}{\mathrm{d}x} = -a \cdot i_r$	(6-22)
Transfer current density	$i_r = i_{0,r} \cdot \left(\exp\left(\alpha_{an,r} \dfrac{F}{RT} \eta_r \right) - \exp\left(-\alpha_{cat,r} \dfrac{F}{RT} \eta_r \right) \right)$	(6-23)
Overpotential	$\eta_r = \varphi_{S,r} - \varphi_{L,r} - E_{eq,r}$	(6-24)
Equilibrium potential	$E_{eq,r} = E_{ref,r} + \dfrac{RT}{F} \ln\left(\prod_{i=1}^{4} c_{surf,r,i}^{v_{r,i}} \right)$	(6-25)
Exchange current density	$i_{0,r} = F k_{0,r} c_{ref} \prod_{i=1}^{4} \left(\dfrac{c_{surf,r,i}}{c_{ref}} \right)^{v_{an,r,i} \cdot \alpha_{cat,r}} \left(\dfrac{c_{surf,r,i}}{c_{ref}} \right)^{v_{cat,r,i} \cdot \alpha_{an,r}}$	(6-26)

Table 6-6: Formulas applied in the model for the description of the mass transport

Type of formula	Formula	
Mass transport equation	$\varepsilon \dfrac{dc_{\text{bulk},r,i}}{dt} = -u_r \cdot \dfrac{dc_{\text{bulk},r,i}}{dy} + \dfrac{a}{F} i_r \cdot \nu_{r,i}$	(6-27)
Inner mass transport	$c_{\text{surf},r,i} = c_{\text{bulk},r,i} + \dfrac{\delta \cdot i_r}{D_i \cdot F} \nu_{r,i}$	(6-28)
Sherwood correlation	$Sh_{i,r} = a_{\text{Sh}} Re^{b_{\text{Sh}}} Sc_{i,r}^{c_{\text{Sh}}}$	(6-29)
Reynolds number	$Re_r = \dfrac{u_r \cdot \rho_{\text{el}} \cdot d_{\text{f}}}{\mu_{\text{el},r}}$	(6-30)
Schmidt number	$Sc_{i,r} = \dfrac{\mu_{\text{el},r}}{D_i \cdot \rho_{\text{el}}}$	(6-31)
Diffusion layer thickness	$\delta_{i,r} = \dfrac{d_{\text{f}}}{Sh_{i,r}}$	(6-32)

Table 6-7: Auxiliary equations for the characterization of the carbon felt electrode and the description of the electrolyte

Type of formula	Formula	
Specific surface area of the electrode	$a = \dfrac{4}{d_{\text{f}}}(1 - \varepsilon)$	(6-33)
Porosity of the electrode	$\varepsilon = 1 - \dfrac{w}{t \cdot \rho_{\text{carbon}}}$	(6-34)
Compression rate of the electrode	$CR = 1 - \dfrac{t}{t_0}$	(6-35)
State of Charge of positive and negative electrolyte	$SoC_{\text{NE}} = \dfrac{c_{\text{in},1}}{c_{\text{in},1} + c_{\text{in},2}}$ $SoC_{\text{PE}} = \dfrac{c_{\text{in},4}}{c_{\text{in},3} + c_{\text{in},4}}$	(6-36)
Electrolyte viscosity	$\mu_{\text{el},r} = \mu_{\text{offset},r} + \mu_{\text{slope},r} \cdot SoC_r$	(6-37)
Electrolyte conductivity	$\kappa_{\text{el},r} = \kappa_{\text{offset},r} + \kappa_{\text{slope},r} \cdot SoC_r$	(6-38)

Boundary conditions that are required to solve the set of equations are listed in table 6-8 and table 6-9, for NE and PE respectively. The effective current density can be achieved by the integral of the solid phase current density at the boundary between bipolar plate and carbon felt electrode for the NE side over height. Liquid phase current density at the interface between electrode and bipolar plate must be zero, whereas the solid phase current density must be zero at the electrode membrane interface. Additionally, no mass transport can occur through the membrane or through the bipolar plate.

Table 6-8: Boundary conditions for the negative electrode

Type of formula	Formula	
Current density distributed over the height of the cell	$j = j_{S,NE}(x = 0, y)$	(6-39)
Current density applied to the cell	$j_{\text{eff}} = \dfrac{1}{H} \displaystyle\int_0^H j(y)\,dy$	(6-40)
Liquid phase current density at the bipolar plate interface of the NE	$j_{L,NE}(x = 0, y) = 0$ for $0 < y < H$	(6-41)
Solid phase current density at the membrane interface of the NE	$j_{S,NE}(x = t, y) = 0$ for $0 < y < H$	(6-42)
Solid phase potential at the bipolar plate interface of the NE	$\varphi_{S,NE}(x = 0, y) = 0 - r_{bpp,NE} \cdot j(y)$ for $0 < y < H$	(6-43)
No mass transport through the bipolar plate of the NE	$\dfrac{dc_{\text{bulk,NE},i}(x = 0, y)}{dx} = 0$ for $0 < y < H$	(6-44)
No mass transport through the membrane from the NE	$\dfrac{dc_{\text{bulk,NE},i}(x = t, y)}{dx} = 0$ for $0 < y < H$	(6-45)
Inlet concentration for the NE	$c_{\text{bulk,NE},i}(x, y = 0) = c_{NE,in,i}$ for $0 < x < t$	(6-46)

Table 6-9: Boundary conditions for the positive electrode

Type of formula	Formula	
Liquid phase potential at the membrane interface of the PE	$\varphi_{L,PE}(x' = 0, y) = \varphi_{L,NE}(x = t, y) - r_{mem} \cdot j(y)$ for $0 < y < H$	(6-47)
Solid phase current density at the membrane interface of the PE	$j_{S,PE}(x' = 0, y) = 0$ for $0 < y < H$	(6-48)
Solid phase current density at the bipolar plate interface of the PE	$j_{S,PE}(x' = t, y) = j(y)$ for $0 < y < H$	(6-49)
Liquid phase current density at the membrane interface of the PE	$j_{L,PE}(x' = 0, y) = j(y)$ for $0 < y < H$	(6-50)
No mass transport through the bipolar plate of the PE	$\dfrac{dc_{bulk,PE,i}(x' = 0, y)}{dx} = 0$ for $0 < y < H$	(6-51)
No mass transport through the membrane from the PE	$\dfrac{dc_{bulk,PE,i}(x' = 0, y)}{dx} = 0$ for $0 < y < H$	(6-52)
Total cell voltage	$U = \varphi_{S,PE}(x' = t, y) - r_{bpp,PE} \cdot j(y)$	(6-53)
Inlet concentration for the PE	$c_{bulk,PE,i}(x', y = 0) = c_{PE,in,i}$ for $0 < x < t$	(6-54)

6.3 Experimental setup for cell & electrolyte characterization

An automated redox-flow battery test stand (FuelCon AG, Germany) as already described in chapter 3 and 5 was used for this investigation. Briefly, the test stand comprises of two temperature controlled electrolyte vessels, that are purged with pure nitrogen to prevent side reactions. Gear pumps and flow meters allow to pump a controlled flow rate through the cell, which is connected to an electrical load and voltage supply, with the possibility to run the cell in galvanostatic or potentiostatic mode and with the ability to acquire full cell impedance spectra. Online conductivity measurements combined with electrolyte sampling and SoC determination by titration with potassium permanganate give precise information on the SoC of both electrolytes during the measurement. A single 10x10 cm² flow through design cell is connected to the test stand. The cell is equipped with four self-made liquid and four self-made solid phase potential probes placed at the electrode-membrane and electrode-bipolar plate interface for each half-cell. The cell comprises untreated GFD4.6EA carbon felt (SGL Carbon, Germany) in both half-cells and a Nafion® 117 membrane (Chemours, USA) separating both electrodes, soaked for at least 24 h in 1% H_2SO_4. The carbon felt was compressed to 2.67 mm, which corresponds to a compression ratio of 42%. A flat PPG86 bipolar plate (Eisenhuth GmbH & Co. KG, Germany) with a thickness of 6.5 mm is contacted to a gold coated nickel mesh and a brass current collector, for NE and PE half-cell, respectively. 30 polarization curves are recorded with alternating voltage steps between 0.6 and 2.0 V, for six different flow rates and five different states of charge. Impedance spectra for similar conditions are recorded prior to the acquisition of the polarization curves. The results for a GFD4.6EA felt with a compression rate of 9% are maintained from the last chapter, but the results of a cell equipped with a similar electrode compressed to a compression ratio of 42% are presented as well.

The electrolyte was purchased from GfE (Gesellschaft für Elektrometallurgie, Germany, 3.9 M sulfate, 1.6 M vanadium, 0.06 M phosphate, <10 mM iron, 6 L on each side). Electrolyte characterization was carried out by discharging stepwise a fully charged vanadium electrolyte and sampling sufficient amounts of NE and PE to determine SoC, density and kinematic viscosity. The SoC was analyzed by titration with potassium permanganate accordingly to chapter 3. The density was determined by weighing 1 mL electrolyte, accurately sampled with an Eppendorf pipette, for NE and PE respectively. Kinematic viscosity was obtained by using and Ubbelohde viscometer, carefully temperature controlled to 25 °C in a water bath. A few samples were analyzed for each SoC, thus allowing the determination of average values and standard deviations.

6.4 Evaluation of electrolyte properties, open circuit cell voltage, kinetic data, membrane and bipolar plate resistance

Prior to the model validation and parameter estimation of mass transport parameters, it is necessary to evaluate all the data that is accessible without the need for a complex mathematical model. These are the viscosity, density and conductivity of the electrolyte, open circuit voltages and resulting reference potentials, membrane resistances for charging and discharging and reaction rate constants.

The electrolyte viscosity and density are given in Fig. 6-4 A). In general, the density of the NE is larger compared to the PE and increases with decreasing SoC, whereas the density of the PE decreases with decreasing SoC. For high flow rates the density is quite similar for NE and PE with an average value of approximately 1350 kg·m⁻³. The relative change for low SoC is less than 2% and the average between NE and PE remains quite constant at 1350 kg·m⁻³. Therefore, the density of the electrolyte is estimated to stay constant at this value independent of SoC or half-cell. The viscosity instead shows a more distinct dependence on SoC, with NE and PE showing both a minimum at high SoC and a quite linear increase for lower

SoC. Linear approximations for both viscosities are given in Fig. 6-4 A). The conductivity data obtained from the online SoC measurement is shown in Fig. 6-4 B), with linear correlations given as well. The NE conductivity is around 8 S·m^{-1} lower than the PE conductivity and both increase with increasing SoC. This corresponds to the data of Corcuera et al. [43].

During the recording of polarization curves a couple of data points were acquired under open circuit conditions and for several SoC, thus allowing an evaluation on open circuit cell voltage dependence on SoC as given in Fig. 6-4 C). Since the SoC of NE and PE were not necessarily identical, the average SoC is used for presenting the data, however for calculation of the open circuit voltage both precisely measured SoCs are available. In general, the Nernst equation is a valid tool to calculate open circuit cell voltages based on thermodynamic considerations. When applying the standard electrode potentials as given in equation 6-11 and 6-12 (c.f. [41]) and calculating the cell open circuit potential E_{OCV} potential based on a simple Nernst equation (c.f. equation 6-25 and 6-57) considering only vanadium species, a quite strong deviation to the measurements of approximately 200 mV can be found, although the curvature of the dependence seems reasonable. Some improvements can be made, when the proton concentration in both electrolytes is taken into account. Therefore, the following two equations of charge balance and dissociation equilibrium must be fulfilled for both electrolytes:

$$0 = \sum_k c_k z_k \tag{6-55}$$

$$K_a = \frac{c_{H^+} \cdot c_{SO_4^{2-}}}{c_{ref} \cdot c_{HSO_4^-}} = 10^{-1.92} \tag{6-56}$$

k is an index for all charged species including four vanadium species, H^+, HSO_4^- and SO_4^{2-}. c_k represents the concentration of each species k and z_k is the elementary charge of each species k (i.e. $z_{H^+}=1$, $z_{HSO4^-}= -1$). K_a is defined as the acid dissociation constant of the first dissociation of sulfuric acid [145]. The Nernst equation extended by the proton concentration as they take part in the PE reaction (c.f. equation 1-2 and 6-58) yields a slightly better curve for SoC higher than 40%, however there is still a remaining deviation of approximately 150 mV. Since not only the redox processes at both electrodes are involved in the entire cell potential, but also Donnan potentials over a cation exchange membrane can arise due to the pH gradient between NE and PE, it is likely that they can influence the entire cell voltage. Knehr et al. implemented this contribution into the Nernst equation (c.f. equation 6-59, [95]), resulting in an even better estimation, but still missing the experimental data by around 60 mV.

$$E_{OCV} = E_{ref,PE} - E_{ref,NE} + \frac{RT}{F} \ln \left(\frac{c_{V2+} c_{V5+}}{c_{V3+} c_{V4+}} \right) \tag{6-57}$$

$$E_{OCV} = E_{ref,PE} - E_{ref,NE} + \frac{RT}{F} \ln \left(\frac{c_{V2+} c_{V5+}}{c_{V3+} c_{V4+}} c_{H^+,PE}^2 \right) \tag{6-58}$$

$$E_{OCV} = E_{ref,PE} - E_{ref,NE} + \frac{RT}{F} \ln \left(\frac{c_{V2+}c_{V5+}}{c_{V3+}c_{V4+}} c_{H^+,PE}^2 \frac{c_{H^+,PE}}{c_{H^+,NE}} \right) \tag{6-59}$$

$$E_{OCV} = E_{ref,PE} - E_{ref,NE} + \frac{RT}{F} \ln \left(\frac{c_{V2+}c_{V5+}}{c_{V3+}c_{V4+}} c_{H^+,PE}c_{H^+,NE} \right) \tag{6-60}$$

Since all the previous estimations deviate from the experimental data quite significantly, another estimation approach was applied. The simple variant of the Nernst equation is combined with a constant offset to the reference potential building $E'_{ref,PE}$ to allow the best representation of experimental data (by least squares method, c.f. equation 6-61). A resulting potential of 1.151 V is achieved for $E'_{ref,PE}$, corresponding to a potential offset of 147 mV. This is in an acceptable agreement to the determined PE reference potential from the kinetic measurement setup presented in chapter 4. There the PE reference potential was determined to 1.12 V, which is roughly 30 mV lower than the current value, but this deviation might be related to the substantial lower vanadium concentration in the kinetic measurement setup, that might affect activity coefficients. For the redox-flow battery model equation 6-61 with its optimized reference potential is used.

$$E_{OCV} = E'_{ref,PE} - E_{ref,NE} + \frac{RT}{F} \ln \left(\frac{c_{V2+}c_{V5+}}{c_{V3+}c_{V4+}} \right) \tag{6-61}$$

The data sets from the polarization curve and impedance spectroscopy measurements of the cell with a higher compression rate of 42% can be evaluated accordingly to the procedure explained in chapter 5. This evaluation includes the determination of the bipolar plate resistance by the solid phase potential probes in contact to the bipolar plate, the determination of the membrane resistance during charge and discharge for several SoC and the identification of the exchange current densities based on an impedance model of Paasch et al. [146]. The evaluated membrane resistances are presented in Fig. 6-5 A) and B), whereas the open circuit resistances show a good agreement between the cell with high and low compression rates. The membrane resistances during charge and discharge deviate more distinct from each other and the membrane resistance for the more intensely compressed electrode shows lower values. However, they agree in their general dependence on SoC. The charge resistance increases with increasing SoC, but the discharge membrane resistance decreases with increasing SoC and remains at a higher level. This behavior fits well to the work of Schafner et al. [147] for the charge resistance and the high SoC region during discharge, but is not in agreement for low SoC and discharging currents. Nonetheless a description of the membrane resistances is necessary to be implemented in the redox-flow battery model. Therefore, both membrane resistances are described by a linear function depending on SoC as indicated in Fig 6-5 B).

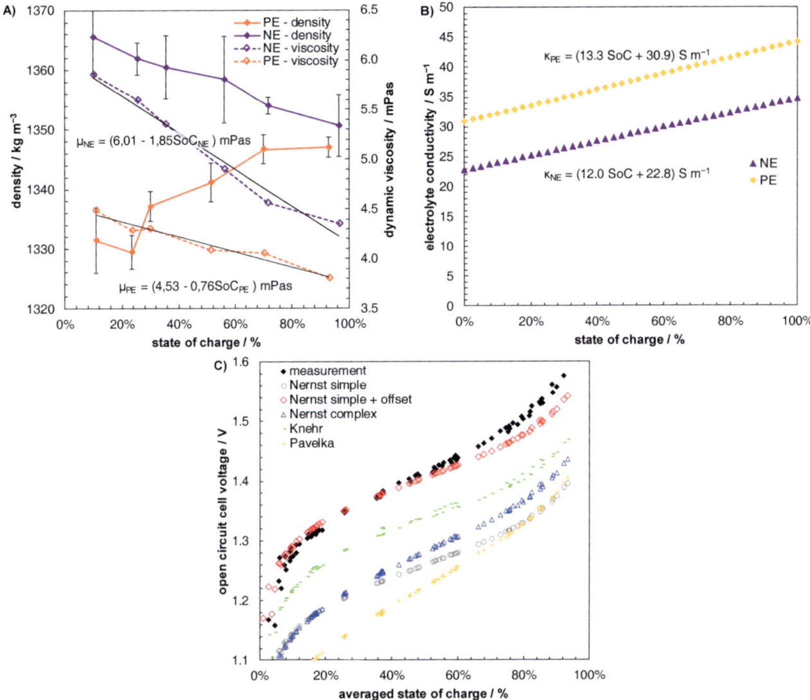

Figure 6-4: A) Measured densities and viscosities for negative and positive electrolyte at different SoC. B) Calculated conductivity depending on SoC based on the initial and post run measurements of conductivity and SoC. C) Comparison of experimental and calculated open circuit cell voltages by using different assumptions [42,95].

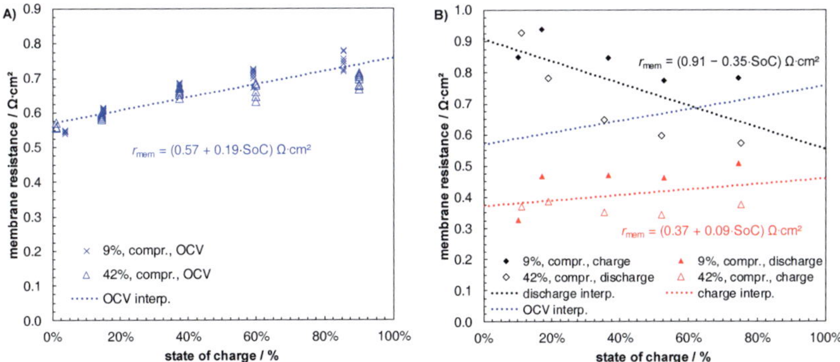

Figure 6-5: A) Membrane resistances under open circuit conditions from EIS-evaluation for two cell setups and linear regression of its dependence on averaged SoC. B) Discharge and charge resistances of the membrane depending on averaged SoC and resulting linear regression lines for charge and discharge resistances.

The evaluation of impedance spectroscopic data by application of the model of Paasch et al. requires an appropriate felt resistance, that can be yielded for an electrode compression rate of 42% from Fig. 2 in chapter 5 and is shown in table 6-10. Accordingly, the carbon fiber diameter is assumed to be 10 µm as well. The evaluation delivers exchange current densities depending on SoC for two compression rates and NE an PE, respectively (c.f. Fig 6-6). The exchange current densities are interpolated based on a reaction order of two, as it gave a good fit in the evaluation part of the last chapter and seems to deliver a more or less appropriate representation of the exchange current density behavior for the 42% compression rate cell. In general, the reaction rate of the 42% compression rate cell is 3 or 4 times higher than for the low compressed cell, which might be caused by normal variations of the material or due to mechanical stress of the more intensively stressed carbon fibers increasing defect sites and improving kinetics. As it was already shown in the last chapter that the exchange current density for the PE half-cell is considerably larger than for the NE half-cell, for both cell setups. The resulting reaction rate constants, felt resistances and bipolar plate resistances are given in table 6-10.

Table 6-10: Reaction rate constants derived from evaluation of electrochemical impedance spectroscopy at open circuit conditions, carbon felt resistance from *ex-situ* measurements, bipolar plate resistance from evaluation of solid phase potential probe signals and calculated porosity of the carbon felt under compression.

	$k_{0,NE}$	$k_{0,PE}$	ρ_S	r_{bpp}	ε
	m s^{-1}	m s^{-1}	Ω m	Ω m^2	%
9% compression rate	$3.0 \cdot 10^{-7}$	$1.6 \cdot 10^{-6}$	$2.2 \cdot 10^{-3}$	$0.9 \cdot 10^{-5}$	94.4
42% compression rate	$8.3 \cdot 10^{-7}$	$5.3 \cdot 10^{-6}$	$1.9 \cdot 10^{-3}$	$0.65 \cdot 10^{-5}$	91.2

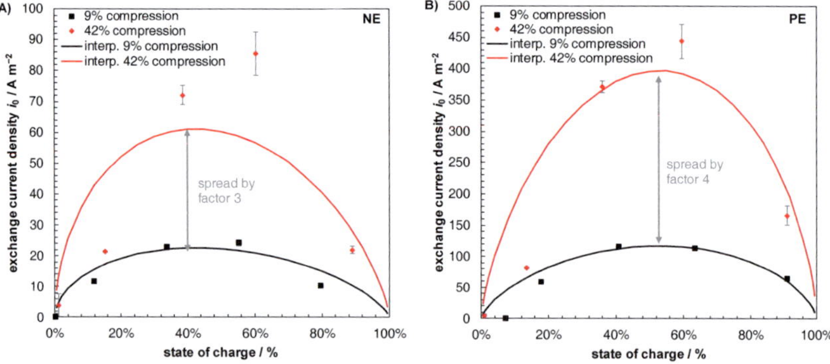

Figure 6-6: A) Averaged membrane resistances of two cell setups with different compression rates of the carbon felts B, C and D) exchange current densities for NE and PE derived from EIS measurements

6.5 Evaluating reasonable diffusion coefficients from literature data

The parameters derived so far in this chapter (reaction rate constant, carbon felt resistance, bipolar plate resistance, membrane resistance, electrolyte conductivity, electrolyte viscosity, open circuit cell voltage), combined with the charge transfer coefficients of chapter 4 form a very good starting point for the mathematical model describing the entire redox-flow cell. The model itself can be applied to determine the mass transport correlation parameter a_{Sh} based on the experimental data of the polarization curves for the two cells equipped with potential probes. However, quite precise diffusion coefficients are obviously crucial to determine the mass transport parameter accurately. For application in mathematical models for vanadium redox-flow batteries, the diffusion coefficients of Yamamura et al. [148] are used quite often (such as but not limited to [41,93,95,101,135,138]). Yamamura et al. determined the diffusion coefficients of all four vanadium species by simulating cyclic voltammograms and estimation of kinetic and mass transport parameters to fit the simulations to the experiments. Additionally, they assumed to have equal diffusion coefficients for the vanadium pairs in positive and negative electrolyte ($D_{V4+} = D_{V5+}$ and $D_{V2+} = D_{V3+}$). The diffusion coefficients they estimated varied between 2.4 and 4.0 $\cdot 10^{-10}$ m²s⁻¹, depending mostly on the type of electrode. As the concentration of vanadium was only 50 mM in a 1 M sulfuric acid solution, it is likely to expect different diffusion coefficients for the typical vanadium electrolyte, with its higher total sulfate concentration (3.9 M) and its higher vanadium concentration (1.6 M) that might affect the diffusion coefficients. Lawton et al. investigated the effect of varying sulfuric acid concentrations on the diffusion coefficient of VO^{2+} by electron paramagnetic resonance and linear sweep voltammetry [149]. They could show a clear decrease of the diffusivity of VO^{2+}, effected by the increase of the sulfuric acid's viscosity and changes of the Stokes radius. The results of Lawton et al., Yamamura et al. and from some more publications is shown in Fig. 6-7. Only two of the shown publications give specific values for the diffusion coefficient of each vanadium species. These are Jiang et al. [150], who calculated the diffusion coefficients by a Car-Parinello molecular dynamics method for an environment in a pure aqueous system and Oriji et al. who employed galvanostatic oxidation or reduction in an electrolyte comprising 5 M sulfuric acid and evaluation by the dependence on transition time and applied current density using Sand's equation [150,151]. The results of both studies show the highest value for the diffusion coefficient of V^{2+} and the lowest value for V^{3+}. VO^{2+} and VO_2^+ are in between and have approximately the same value. The values of Oriji et al. are clearly below those of Jiang et al., most likely due to the higher sulfuric acid concentration, as already emerged from the work of Lawton et al. If the values between these two studies are interpolated linearly, the resulting line for VO^{2+} runs through the range of the dataset of Yamamura et al. and also corresponds to some extent with the value of Gattrell et al. [152]. However, a clear deviation remains to the result of Lawton et al., but due to the need to define diffusion coefficients for all four vanadium species and since the majority of the publications results in values lower than those of Lawton et al. the interpolation as shown in Fig. 6-7 is expected to be the most reasonable one. The interpolated values for a sulfuric acid concentration of 4 M are indicated as well and vary between 0.6·10^{-10} m²s⁻¹ for V^{3+} and 2.3·10^{-10} m²s⁻¹ for V^{2+}. Any additional effect due to higher vanadium concentrations is neglected. The resulting Schmidt numbers are presented in table 6-11 indicating quite high values for the vanadium electrolyte, when compared to the typical Schmidt numbers given for example in table 6-1 and 6-2. In particular, the Schmidt number for V^{3+} species is extremely high underlines the need to integrate Schmidt number dependence into mass transport correlations for vanadium redox-flow batteries.

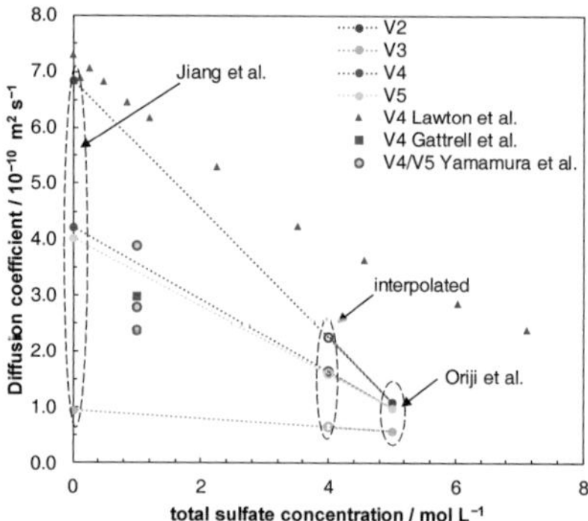

Figure 6-7: Diffusion coefficients of four vanadium species for different total sulfate concentrations based on several publications [148–152] (data of Lawton et al. is reduced to the published data received only by electron paramagnetic resonance) and interpolated values for 4 M total sulfate concentration based on the marked data sets.

Table 6-11: Diffusion coefficients and exemplary Schmidt numbers for electrolyte containing only one vanadium species.

Vanadium species	Diffusion coefficient $m^2\,s^{-1}$	Schmidt number dimensionless
V^{2+}	$2.3 \cdot 10^{-10}$	13398
V^{3+}	$0.6 \cdot 10^{-10}$	74198
VO^{2+}	$1.6 \cdot 10^{-10}$	20972
VO_2^{+}	$1.6 \cdot 10^{-10}$	17454

6.6 Mass transport parameter estimation & model validation

The mathematical model as described in section 6.2 was implemented in gPROMS® ModelBuilder 5.1.1 (Process Systems Enterprise Limited, UK). The grid size of the electrodes play a critical role to accomplish a robust and efficient model. Several model evolutions lead to a robust and quite fast model with grid distributions and discretization methods as follows: The y-dimension (parallel to flow) is separated in 50 grid points with a uniform distribution over the height of the electrode and solved with a second order backward finite difference method. The x-dimension (through plane) required a manually designed non-uniform grid. The mid-section is designed with a coarse grid and both outer regions possess an exponentially decreasing step size for each grid point, resulting in a symmetric grid of 20 intervals, as presented in Fig. 6-8 B). This design is required, since large potential, current density and concentration gradients occur especially at the boundary layer between carbon felt electrode and membrane affecting the stability of the whole model. The discretization method in this dimension is set as centered finite difference with an order of two.

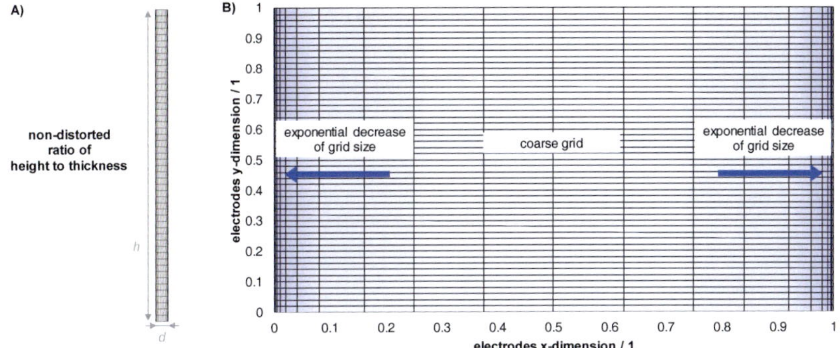

Figure 6-8: A) Sketch of the grid distribution within the electrode with correct ratio of electrode height to electrode thickness for a compression rate of 9%. B) Distorted and dimensionless representation of the grid within the electrode to illustrate the non-uniform grid size in the through-plane direction of the electrode.

The experimental data of all polarization curves (five different SoC each with six times variation of flow rate) resulting from the two cell setups of 9% and 42% carbon felt compression rate was implemented in gPROMS, so that quasi-stationary current density and potential probe signals from modelling could be compared to experimental data, when operational parameters (flow rate, cell voltage and state of charge of each electrolyte) were assigned in the model. The potential probe signals could be directly extracted from the model data of liquid and solid phase potential distribution, when the corresponding positions of the experimental probes were used (c.f. appendix A.2, full cell model). Other parameters, such as reaction rate constants, membrane resistances, reference potentials, electrolyte properties, specific electrical resistance of the carbon felt were taken from the evaluation of previous sections and chapters. The model validation capability of gPROMS® was executed to determine the mass transport parameter a_{Sh} by the maximum likelihood estimation method for both cell setups individually (all involved solvers and solver parameters were kept at the default value).

The results of the parameter estimation of the parameter a_{Sh} are given in Fig. 6-9 A) and is compared to those of Kinoshita et al. and Vatistas et al. [128,141]. The results are in the range of these two publications, but are closer to the values of Kinoshita et al. Additionally, the corresponding Sherwood numbers for a given Schmidt number of completely discharged positive electrolyte are compared to those from Kinoshita et al., Vatistas et al. and Schmal et al. [94,128,141], whereas their experimentally derived data points are presented in order to compare the accuracy of the methods (c.f. Fig. 6-9 B). The average correlation from the two parameter estimations fits quite well to the datasets Kinoshita et al. presented, especially when considering the range of uncertainty from both evaluations. The dataset of Vatistas et al. is in fairly good congruence only for low Reynolds numbers, however they mentioned a higher uncertainty for their results in this range. The range of Sherwood numbers between those predicted by Kinoshita et al. and those predicted by Vatistas et al., was assumed to be the most reasonable one (c.f. section 6.1), thus it seems that the parameter estimation confirms this assumption giving a mass transport correlation as follows:

$$Sh_i = 0.07 \pm 0.02 \, Re^{0.66} Sc_i^{0.45} \qquad (0.0018 < Re < 0.11) \qquad (6\text{-}62)$$

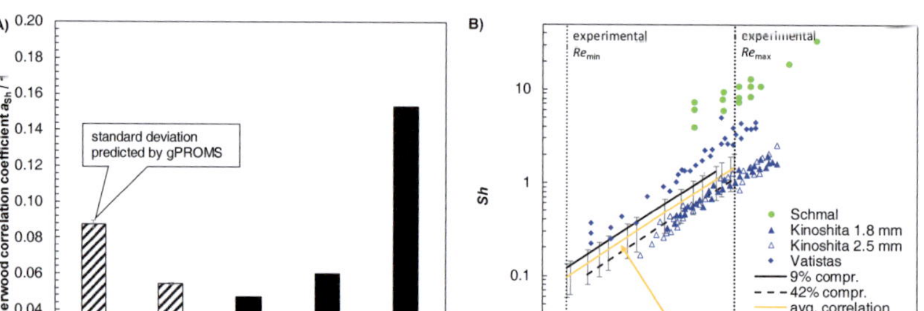

Figure 6-9: A) Resulting Sherwood correlation coefficients a_{Sh} of both experimental setups compared to the values of Kinoshita and Vatistas et al. [128,141]. B) Sherwood-Reynolds dependencies exemplary shown for a Schmidt number of 20972 (fully discharged PE). The black lines are the correlations resulting from parameter estimation, whereas the yellow line represents the average of both estimations. To achieve a better comparison between the results from parameter estimation and the results of Schmal et al., Vatistas et al. and Kinoshita et al. their datasets of the experimentally accessible data points is presented instead of the resulting correlation.

To check the accuracy of the model predictions for current density and potential probe signals, a subset of characteristic curves is shown in Fig 6-10 to Fig 6-13. The full data set can be found in the appendix (c.f. section A.1), where experimental data and modelled values are compared in their entirety, but for the sake of readability, the presented results are reduced to the most characteristic ones in this section. Fig 6-10 presents the comparison of polarization curves for variation of flow rate at low states of charge for both cell setups. The low state of charge results in a sufficient mass transport limitation for low flow rates, where a distinct maximum current density can be achieved during discharge. Modelled and experimentally derived curves are generally in an acceptable agreement and in particular the polarization curves for low and high flow rates show a reasonable well congruence during discharge. Slight deviations occur during the charging process, where the modelled data underestimate the current density and during the discharge process for medium flow rates, where the current density is slightly overestimated by the model. A possible explanation could base on additional side reactions during the charging process, in particular the hydrogen evolution reaction in the negative half-cell, which could lead to higher charging current densities. Since the model is not designed to describe side reactions, this underestimation of the current density during charging could cause the model to increase the mass transport coefficient, leading to an overestimation of the current density during discharge. At very low and very high flow rates this effect is less important because very low flow rates limit the discharging current density through strong convective mass transport limitations, superimposing the weaker pronounced effects of diffusive mass transport. At high flow rates the convective mass transport is obviously not limited and diffusion mass transport limitation becomes negligibly small, since the diffusion boundary layer is decreased to low thicknesses. Only at medium flow rates does the diffusion mass transport influence the discharging current more intensively and thus provide a potentially good explanation for the observable deviation.

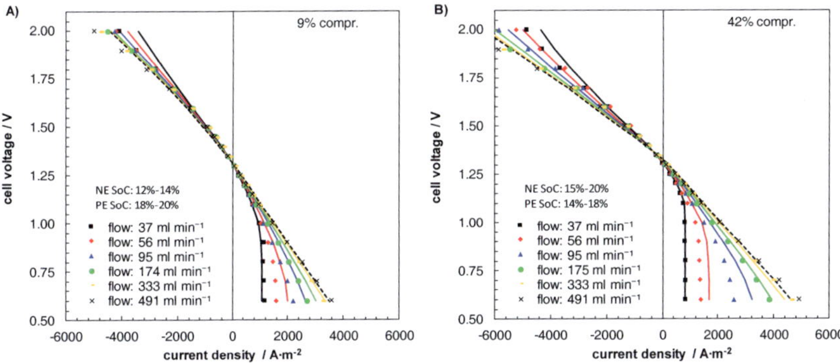

Figure 6-10: A) Polarization curves for variation of flow rate at low states of charge for the cell setup with a compression rate of 9%. The lines represent modeled data and data points indicate the experimental data. B) Polarization curves for variation of flow rate at low states of charge for the cell setup with a compression rate of 42%.

When considering the results presented in Fig. 6-11, where polarization curves for low and high flow rates under variation of state of charge are presented for both cell setups, it seems reasonable to assume a certain degree of side reactions during charging, since the current density during charging is underestimated by the model especially for low flow rates. For higher flow rates, the current densities increase in general due to increased mass transport thus the deviation due to potential side reactions is mitigated. Nevertheless, the predicted curves are in a quite good agreement to the measured polarization curves for all conditions shown, at least for the cell voltage as the key indicator for the cell's performance.

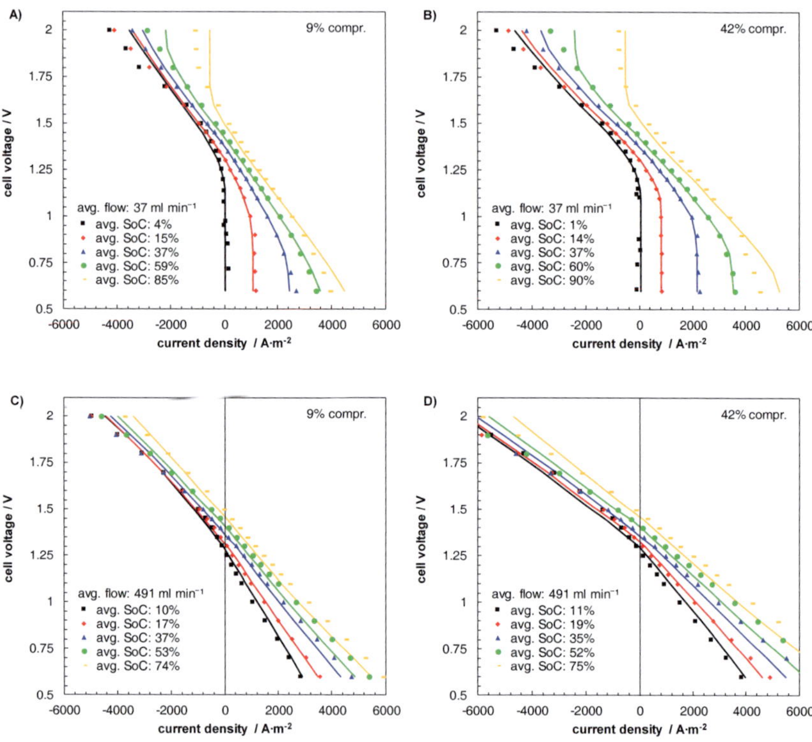

Figure 6-11: Polarization curves of both cell setups for compression rates of 9% and 42% under variation of state of charge. A) 9% carbon felt compression rate for low flow rates. B) 42% carbon felt compression rate for low flow rates. C) 9% carbon felt compression rate for high flow rates. B) 42% carbon felt compression rate for high flow rates. Data points represent experimental data, lines indicate data from model validation.

As already presented in chapter 5 Fig. 4, the difference between liquid and solid phase potentials give direct information on the locally resolved overpotential in the VRFB. To capture the whole image of operational parameters in a few figures, the overpotential information is given for low and high flow rates as well as for low and high state of charge in Fig. 6-12. The overpotential is shown for the membrane electrode interface and the bipolar plate electrode interface of the cell setup with 42% carbon felt compression rate for the negative and positive half-cell, respectively. The results indicate in general a good fit between modelled and experimental data, especially the large spread between the high overpotentials at the membrane interface compared to the low overpotentials at the bipolar plate interface. The "potential jumps" at low flow rates that occur at low state of charge during discharge and at high state of charge during charge are predicted quite well, although the maximum current density during charge is underestimated slightly.

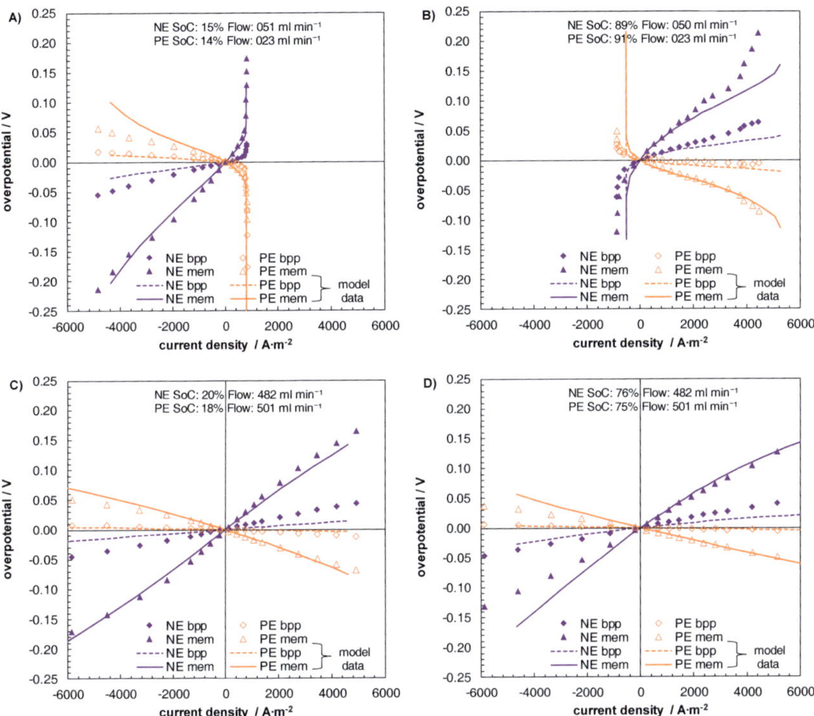

Figure 6-12: Exemplary overpotentials for the positive and negative half-cell of the cell setup with 42% carbon felt compression rate measured at the membrane electrode interface (mem) and at the bipolar plate electrode interface (bpp) under variation of flow rate and SoC. A) low flow rate and low SoC B) low flow rate and high SoC C) high flow rate and low SoC D) high flow rate and high SoC. Points represent experimental data, whereas lines indicate results from the mathematical model.

The curve of the overvoltage signals at the interface to the membrane is mainly determined by the curve of the liquid phase probe signal. Due to the significantly lower specific electrical resistances in the felt material, the solid phase signals are only very faintly developed and result in an almost always linear curve (the reader might be referred to the appendix section A.1 where all polarization curve data of cell voltage and potential probe signals is included). The signals at the bipolar plate interface typically possess only a very low inclination as well, so only the curve of liquid phase probes at the membrane interface with variation of flow rate at low state of charge is shown in Figure 6-13. The low state of charge accentuates mass transport limitations during discharge, while the signals for charging show a very narrow range for flow rate variations. The comparison of experimental and modelled data during discharge indicates an at least sufficient congruence. However, the cell setup with a carbon felt compression rate of 9% shows distinct deviations, in particular for the negative half-cell, where the slope for charging seems to be mismatched and the liquid phase potential during discharging is slightly overestimated. This is somehow surprising, since the exchange current densities from impedance spectroscopy showed a quite good fit for the negative half-cell, when compared to the positive half cell (c.f. Fig. 6-6). Side reactions are unlikely to cause the deviation in overpotentials during charging, since the anodic overpotential in the negative half-cell rather suppresses hydrogen evolution reaction than promoting it. Therefore, the reason for the deviation might be related to uncertainties in the charge transfer coefficients, that showed a slight dependence on state of charge as indicated in chapter 4 Fig. 10. Nonetheless, this is still not a very probable explanation, since it cannot elucidate the quite good fit for the cell setup with higher carbon felt compression rate (Fig. 6-13 B). At least the tendencies of the probe potentials are predicted quite well for all liquid phase potential probe signals shown in Fig. 6-13. Although the figures are not given in this section, it should be mentioned that the model is quite precise in predicting the solid phase potential probe signals for almost every operational condition and cell setup. The reader is referred to the appendix, where a comparison of all experimental and modelled data is given.

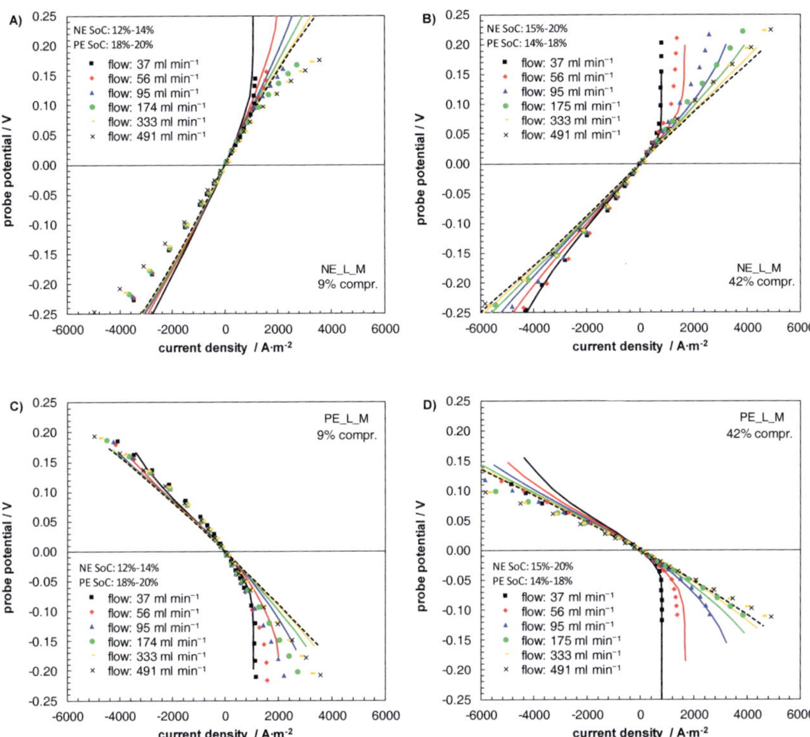

Figure 6-13: Liquid phase potential probe signals at carbon felt membrane interface for low state of charge under variation of flow rate for positive and negative half-cell, respectively. A) Liquid phase potential in the negative half-cell for the cell setup with 9% carbon felt compression rate B) Liquid phase potential in the negative half-cell for the cell setup with 42% carbon felt compression rate. C) and D) As in A)&B), but for the liquid phase potential in the positive half-cell. Data points represent experimental data, lines indicate data from model validation.

6.7 Conclusion

In this chapter, it was shown how a mathematical model can be applied for parameter estimation of critical mass transport coefficients. To achieve this, the available literature was briefly reviewed to identify a basis for the mass transport correlation and to reduce the number of parameters that needed revision, to just one the Sherwood correlation coefficient a_{Sh}. The most probable value for this parameter was assumed to be in the range of 0.05–0.15. A two-dimensional model was designed based on literature data to describe the electrode processes and all the required input parameters were taken from the previous chapters or were determined experimentally *in-situ* or *ex-situ*. The combination of liquid phase and solid phase potential probes was used during polarization curve measurements to acquire information on the bipolar plate resistance including contact resistance, as well as membrane resistance for charging and discharging and depending on SoC. The reaction rate constants were evaluated from impedance spectroscopy as shown in the previous chapter, while a reaction order of two was assumed for each reacting vanadium species. The conductivity measurements and open circuit cell voltage measurements during cell operation served as data basis to estimate electrolytes conductivity and reference potentials of the positive half-cell. Additional measurements of electrolyte viscosity and density were a reliable source to describe these properties of the electrolyte. A brief literature review on diffusion coefficients for vanadium species in sulfuric acid environment allowed an interpolation of diffusion coefficient of each species for the sulfuric acid concentration applied in the experiments. The parameter estimation application of gPROMS® ModelBuilder was used to estimate the Sherwood correlation coefficient a_{Sh} based on the polarization curves of two cell setups with 9% and 42% carbon felt compression rate. The resulting coefficients are 0.09 and 0.05 and lie well within the expected range, leading to an average value of 0.07. The comparison of modelled data to experimental data exhibits a good congruence although some deviations still remain, which might be explained by side reactions, in particular hydrogen evolution reaction during charging, and uncertainties in the exact description of charge transfer coefficients and exchange current densities. Nevertheless, the model seems to be sufficiently valid and can therefore be used in optimization approaches for the optimization of electrode thicknesses, optimal flow rates or it is applied to predict the effect of electrode treatment, adapted carbon fiber diameters or effects of electrolyte maldistribution due to improperly designed flow fields.

6.8 Symbols and Subscripts applied in the model

Table 6-11: Symbols applied in the mathematical model

Symbol	Name	Unit
a	Specific surface area of the electrode	$m^2\ m^{-3}$
a_{Sh}, b_{Sh}, c_{Sh}	Empirical coefficients for the calculation of the Sherwood number	-
c_{in}	Concentration of vanadium species in cell inlet	$mol\ m^{-3}$
c_{bulk}	Concentration in the bulk of the electrolyte	$mol\ m^{-3}$
CR	Compression rate of the electrode	%
c_{ref}	Reference concentration according to standard conditions	$1000\ mol\ m^{-3}$
c_{surf}	Concentration on the fibre surface of the carbon felt electrode	$mol\ m^{-3}$
D	Diffusion coefficient	$m^2\ s^{-1}$
d_f	Diameter of the graphite fibres	m
d_h	Hydraulic diameter of the carbon felt	m
E_{eq}	Equilibrium potential of a single reaction	V
E_{OCV}	Open circuit potential of a full cell	V
E_{ref}	Standard electrode potential	V
F	Faradaic constant (96485)	$96485\ C\ mol^{-1}$
H	Height of the electrode	m
i	Charge transfer current density	$A\ m^{-2}$
i_0	Exchange current density	$A\ m^{-2}$
J	Mass flux	$mol\ m^{-2}\ s^{-1}$
j	Local current density	$A\ m^{-2}$
j_{eff}	Effective current density applied to the whole cell	$A\ m^{-2}$
j_L	Current density in the liquid phase of the electrode area	$A\ m^{-2}$
j_S	Current density in the solid phase of the electrode	$A\ m^{-2}$
k_0	Reaction rate constant	$m\ s^{-1}$

Symbol	Name	Unit
K_s	Acid dissociation constant	-
k_m	Mass transfer coefficient	$\mathrm{m\ s^{-1}}$
R	Ideal gas constant	$8.314\ \mathrm{J\ mol^{-1}\ K^{-1}}$
r_{bpp}	Bipolar plate resistance	$\Omega{\cdot}\mathrm{m^2}$
Re	Reynolds number	-
$r_{mem,ch}$	Membrane resistance during charging	$\Omega{\cdot}\mathrm{m^2}$
$r_{mem,dis}$	Membrane resistance during discharging	$\Omega{\cdot}\mathrm{m^2}$
Sh	Sherwood number	-
SoC	State of Charge	%
T	Temperature	K
t	Thickness of the compressed electrode	m
t_0	Uncompressed thickness of the electrode	m
u	Electrolyte flow velocity	$\mathrm{m\ s^{-1}}$
U	Cell voltage	V
$\dot{V}$	Flow rate through the electrode	$\mathrm{m^3\ s^{-1}}$
W	Width of the electrode	m
w	Areal weight of the electrode	$\mathrm{kg\ m^{-2}}$
x	Horizontal dimension in the NE electrode	m
x'	Horizontal dimension in the PE electrode	m
y	Vertical dimension	m
z	Elementary charge	-
α_{an}	Anodic charge transfer coefficient	-
α_{cat}	Cathodic charge transfer coefficient	-
δ	Diffusion layer thickness	m
ε	Porosity of the electrode	%
κ_{el}	Electrolyte conductivity	$\mathrm{S\ m^{-1}}$
κ_{offest}	Electrolyte conductivity, offset parameter	$\mathrm{S\ m^{-1}}$
κ_{slope}	Electrolyte conductivity, slope parameter	$\mathrm{S\ m^{-1}}$
μ_{el}	Electrolyte viscosity	Pa s
μ_{offset}	Electrolyte viscosity, offset parameter	Pa s

Symbol	Name	Unit
μ_{slope}	Electrolyte viscosity, slope parameter	Pa s
ν	Stoichiometric coefficient	-
ν_{an}	Reaction order for the anodic reaction	-
ν_{cat}	Reaction order for the cathodic reaction	-
ρ_{carbon}	Carbon fiber density of the carbon felt electrode	kg m^{-3}
ρ_L	Specific electrical resistance of the liquid phase	Ω m
ρ_S	Specific electrical resistance of the solid phase / carbon felt electrode	Ω m
ρ_{el}	Density of the electrolyte	kg m^{-3}
φ_L	Potential in the liquid phase of the electrode	V
φ_S	Potential in the solid phase of the electrode	V

Table 6-12: Indexes and Subscript applied in the description of the mathematical model

Index/Subscript	Name
bulk	Variable or parameter related to the bulk phase of the electrolyte in the electrode
el	Electrolyte parameter
i	Index of the vanadium species
in	Inlet conditions
k	Index of acid species
L	Liquid phase
NE	Negative electrolyte
offset	Offset value of a parameter for a SoC of 0%
PE	Positive electrolyte
r	Index of the reaction
S	Solid phase
slope	Slope of a parameter in dependence of SoC
surf	Variable or parameter related to the surface of the carbon fibers

7 Concluding discussion & outlook

In this work the validation of a two-dimensional numerical model describing the behavior of a vanadium redox-flow battery based on polarization curve measurements was carried out. Membrane crossover effects were neglected in the model, since the time scale of polarization curve measurements is too short and the membrane crossover processes are too slow to have a significant impact on electrolyte imbalance and thus cell voltage. Therefore, the main objectives were a correct description of electrode and membrane potentials within the cell. Carbon fiber based potential probes for the liquid and solid phase were developed to measure liquid and solid phase potentials during the recording of the polarization curves. The resulting measurements built the reference data set the model should predict most precisely. However, a couple of parameters, such as specific electrical resistance of the carbon felt, area specific resistance of the membrane, reaction rate constants, charge transfer coefficients or practical reference potentials of the underlying electrochemical reactions would need adaption by the model to fit the measured data set. This however is a challenging and time consuming task for the modelling environment, potentially delivering uncertain or unrealistic results. To reduce the parameters needing model assisted parameter estimation, a couple of experimental investigations and preliminary evaluations of the polarization curve data sets were necessary to generate own data. These own data are advantageous for model validation compared to data from the literature, as uncertainties or variations in experimental methods in the determination of parameters (other storage or pretreatment of the components, unnamed additives or unnoticed impurities in the electrolyte or membrane) can be excluded.

Preliminary determination of parameters

The charge transfer coefficients were determined in an experimental setup, which provided superior mass transport due to very high flow rates. Additionally, electrode thickness and resulting data were checked by a mathematical model to ensure a homogeneous transfer current density. This enabled a sufficient certainty to precisely measure charge transfer coefficients. The findings indicate charge transfer coefficients, that are in a range of 0.24 to 0.36, thus significantly below the values usually applied in mathematical modelling (typically charge transfer coefficients are fixed to values very close to 0.5). However, the applied electrodes were not exactly the same as applied in the cell, since the quite thick material in the cell setup (about 4.6 mm uncompressed), would lead to major gradients in the transfer current density, thus it had to be omitted for experimental investigations in this setup. Since the thin electrodes (about 0.2 to 0.4 mm uncompressed) had comparable charge transfer coefficients although one was a carbon felt and the other a carbon paper electrode, it is very likely that the thick electrode, being a carbon felt electrode as well, should possess comparable charge transfer coefficients.

The reaction rate constants of the carbon felt electrodes used in the cells were determined via electrochemical impedance spectroscopy of a full cell setup by using an analytical model of Paasch et al. The resulting dependence of exchange current densities on state of charge indicated a reaction order higher than one as probable and the assumption of a reaction order of two for each reacting species in anodic and cathodic reactions resulted in a good agreement to the experiments, for NE and PE respectively.

The membrane resistances were investigated by the evaluation of liquid phase potential probe signals, whereas the probes were placed in direct contact to the membrane at the membrane electrode interface. High flow rates maintained a homogenous current density across the cell height (confirmed by the solid phase potential probe measurements in chapter 3), thus the measured membrane polarization curves are a valid representation of the membrane potential drop for different current densities and were determined for a variation of state of charge. The membrane polarization curves were found to be quite linearly depending on current density, thus it could be well described only by area specific resistances without any additional potential gradients. The membrane resistance during charging is significantly lower than during discharging and the charging resistance increases slightly with state of charge, whereas the discharging resistance decreases more distinctly with increasing state of charge. To implement this behavior into the mathematical model the area specific membrane resistance was described by linear dependence on state of charge for charge and discharge resistance respectively.

Quite comparable to the evaluation of membrane resistances, the area specific resistance of the bipolar plate was determined. However, the bipolar plate resistance is, as expected, unaffected by current density or state of charge. Other parameters such as specific electrical resistance of the felt or electrolyte parameters as density or viscosity were examined in simple *ex-situ* measurements. Only the diffusion coefficients of the four vanadium species had to be extracted from literature data by interpolation to an appropriate total sulfate concentration. The implementation of all these parameters into the model reduced the number of uncertain parameters and just the parameters for mass transport from the bulk phase to the carbon fiber surface remained undetermined. A brief literature overview left just one parameter uncertain that needed parameter estimation by the mathematical model.

Parameter estimation and model validation

The parameter estimation for data sets of two cell setups with different carbon felt compression rates resulted in two quite close values for the coefficient a_{Sh} in the Sherwood correlation, that were averaged to a value of 0.07. The comparison of measured and predicted values of the model indicate in general a good congruence between the experiments and the model. Therefore, the model seems to be valid and the resulting mass transport correlation narrows the range of mass transport correlation available in literature. Of course, it has to be considered that the diffusion coefficients were interpolated and that new investigations might find diffusion coefficients in vanadium electrolyte that differ to the interpolated values used in this work, thus a recalculation of the mass transport coefficient might become necessary. When comparing modelled and experimental data especially the liquid phase potential probe signals show a certain deviation. This probably attributes to some inaccuracies in the description of kinetics and the membrane potential drops. The electrode is assumed to possess isotropic properties, however the distribution of carbon fibers within the electrode is somehow stochastic and the carbon fibers within might undergo very slightly different conditions during production of the felt, resulting in differing electrochemical properties. Furthermore, the general assumption of a reaction order of two needs further confirmation or refinement, thus a precise kinetic characterization of a sufficient amount of single carbon fibers extracted from the felt is highly desirable.

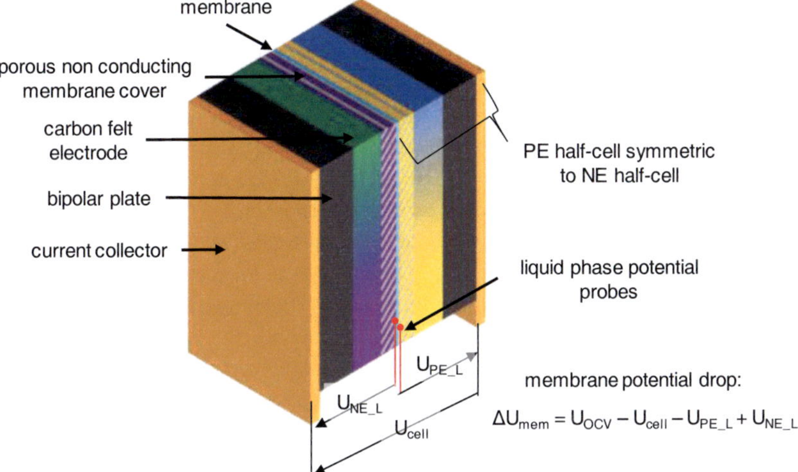

Figure 7-1: Suggestion for a cell setup to measure the membrane voltage drop more precisely. Two liquid phase potential probes attached to the membrane in the NE and PE half-cell are shown, but more couples on several spots of the membrane could increase the validity of the measurement.

Additional investigations in measuring the membrane polarization in an operating cell are another highly important topic, that could be intensified by using liquid phase potential probes in a modified cell setup as shown in Fig. 7-1. The proposed setup bases principally on the same setup as described in this thesis, however the membrane would additionally be covered on both sides by a porous electrically non-conductive material. This cover could be for example a glass fiber felt, ideally possessing a permeability comparable to the carbon felt electrode. It would maintain an almost constant electrolyte composition as in the cell inlet, thus liquid phase potential probes would see always the same redox-potential allowing a precise voltage measurement. A sufficiently high flow rate would ensure a homogenic current density distribution, thus a membrane voltage drop would be accessible in the same way as done in chapter 6. However, multiple probes could be inserted allowing parallel recording of membrane voltage drops at several spots on the membrane and with a probably higher precision, since the electrode reactions that could affect the liquid phase probes would be locally separated. Compared to methods applying impedance spectroscopy, this setup allows direct measurement of polarization curves under real operating conditions without a need for evaluation of impedance data and without integrating differential resistances. Consequently, a direct application of the resulting data into the model should be possible.

Use of the model for optimization of a VRFB

As depicted in the introduction part of this thesis a mathematical model of vanadium redox-flow battery needs validation to enable an analysis of limiting components, to check the sensitivity to different parameters of the components and to improve the understanding of the whole system. One important aspect of this is a proper choice of operational conditions, since the model is preferably applied for cells in commercial scale and not for laboratory scale cells. One essential parameter describing the flow conditions within a cell is the Reynolds number. A typical commercial cell with a cell height of 0.3 to 0.6 m, an electrode thickness of about 1 mm, an electrolyte utilization of 10% per cell passage, an electrolyte viscosity of approximately 5 mPas, a carbon fiber diameter of 10 μm and an applied current density of 1000 A m^{-2} would have a Reynolds range of 0.05 to 0.11, perfectly matching in the experimental range of this work (0.0018 to 0.11, c.f. equation 6-62). Even when applying thinner electrodes, such as single layer carbon paper electrodes, with a thickness of only 0.2 mm yielding higher current densities of 2000 A m^{-2}, with an interdigitated flow field and a distance from inlet to outlet channel of approximately 15 mm would achieve a Reynolds number of 0.026 matching into the experimental range as well.

Therefore, the model should give a very good indication for current density, potential and concentration distributions within an operating cell of commercial scale, helping to understand the limiting processes within the cell. Besides this one could use the model, to predict the effect of several parameter variations of the cell. The most suitable ones would be the variation of cell thickness, reaction rate constants (corresponding to different pretreatment methods) and carbon felt fiber diameters. An optimization of flow rate can be carried out, but additional interdependencies on pressure drop across the cell and the whole piping system would be necessary to find a global optimum and not just the expectable result, that a steadily increasing flow rate will continuously increase the performance correspondingly. A probably more interesting approach would be a simulation of two cells, with differing size and area specific flow rate, but otherwise identical parameters connected in parallel, thus maintaining equal cell voltage. Such a model setup would theoretically represent a cell with uneven flow distribution and the variation of flow rate and cell size of these two cells would allow an analysis and determination of critical values of electrolyte maldistribution and performance loss compared to a system with ideal flow distribution.

Enhancements for the model

The applied model in this work is capable of describing electrode processes within the two electrodes of a redox-flow battery and the polarization of the membrane, thus giving good predictions for the cell voltage under various operational conditions. The model assumes isothermal behavior and does not take into account vanadium crossover processes across the membrane, as well as side reactions. Since charge discharge cycling under constant current or constant power conditions is a typical testing procedure to investigate the crossover processes of a vanadium redox-flow battery, it would be necessary, to add a model that could predict crossover processes from diffusion, migration, electro-osmosis and convection. Such a model was presented and validated as a zero-dimensional model by Schafner et al. [64]. Since the vanadium concentration gradient within a cell during charge and discharge cycling would be rather low, a zero-dimensional model would probably be sufficient. However, also a one-dimensional model distributed over the height of a cell would require only slightly more computation time, but could also predict the membrane crossover in the case of strong concentration gradients along the membrane surface. Important would be the separation of both models, whereas the crossover model would only use the concentration and current density distribution along the membrane but would not directly feed the resulting mass fluxes back into the two-dimensional model for electrode processes. A direct feedback loop between both models will probably affect the stability and robustness of the model negatively, but the improvement in accuracy will presumably be modest, because the crossover leads typically only to losses of a few percent in current efficiency. The mass fluxes predicted by the crossover model would, of course, need implementation in the total mass flux leaving the negative and positive half-cell to predict concentration changes within the tanks. The tanks then feed the cell and therefore the crossover model would have an indirect effect on the two-dimensional cell performance model. The whole model setup is indicated in Fig. 7-2. Additional features of this model setup would be the calculation of electrolyte conductivity and viscosity based on the composition of the electrolyte, so these parameters could then be used in the performance model. Of course, this tank model requires just a zero-dimensional setup, since the tanks are presumably represented by continuously stirred tank reactors very well. Such a model setup would have the pleasant characteristic to be able to describe complete charge discharge cycles and the shift of the vanadium concentration would be observable in the change of the cell voltage as well as in the change of the electrolyte conductivity and thus an experimental validation of the model would be relatively easy to carry out.

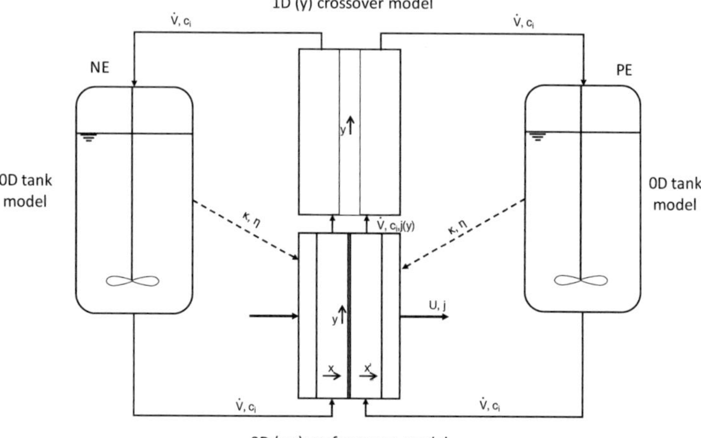

Figure 7-2: Suggestion for a model setup considering a two-dimensional model for the cell performance, a one-dimensional model predicting crossover and a tank model that combines the mass fluxes of both models to calculate the current electrolyte composition and electrolyte properties.

Finally, a last suggestion for the enhancement of the model could be the implementation of non-isothermal behavior. Whereas in a first step only the heat generated by voltage losses, entropy changes during the electrochemical reaction (both predicted by the performance model) and self-discharge (predicted by the cross-over model) is considered to estimate the electrolyte outlet temperature by a straightforward zero-dimensional mass balance. This would allow a validation by the measurements carried out in this thesis, since temperature probes were also installed in the inlet and outlet manifolds of the cell. If the temperature changes would be too distinct, the effect of temperature changes on electrolyte conductivities and viscosities, electrode kinetics and membrane resistance would have to be considered as well. This would, of course, require significantly more experimental investigations to produce a reliable and valid model, but may not really be necessary as the temperature window for vanadium redox-flow batteries is rather limited.

References

[1]   Paris Agreement, United Nations, 2015. https://treaties.un.org/doc/Treaties/2016 /02/20160215%2006-03%20PM/Ch_XXVII-7-d.pdf (accessed December 1, 2019).

[2]   J. Rockström, O. Gaffney, J. Rogelj, M. Meinshausen, N. Nakicenovic, H.J. Schellnhuber, A roadmap for rapid decarbonization, Science. 355 (2017) 1269–1271. https://doi.org/10.1126/science.aah3443.

[3]   Renewable Power Generation Costs in 2018, International Renewable Energy Agency, Abu Dhabi, 2019. https://www.irena.org/-/media/Files/IRENA/Agency/Publication/2019 /May/IRENA_Renewable-Power-Generations-Costs-in-2018.pdf (accessed November 28, 2019).

[4]   H.-M. Henning, A. Palzer, Was kostet die Energiewende? Wege zur Transformation des deutschen Energiesystems bis 2050, Fraunhofer-Institut für Solare Energiesysteme ISE, Freiburg, 2015. https://www.ise.fraunhofer.de/content/dam/ise/de/documents /publications/studies/Fraunhofer-ISE-Studie-Was-kostet-die-Energiewende.pdf (accessed November 16, 2019).

[5]   Infrastructure Outlook 2050, Gasunie Deutschland GmbH & Co. KG and Tennet Holding B.V., 2019. https://www.tennet.eu/fileadmin/user_upload/Company/News/Dutch /2019/Infrastructure_Outlook_2050_appendices_190214.pdf (accessed November 30, 2019).

[6]   M. Robinius, P. Markewitz, P. Lopion, F. Kullmann, P.-M. Heuser, K. Syranidis, S. Cerniauskas, M. Reuß, S. Ryberg, L. Kotzur, D. Caglayan, L. Welder, J. Linßen, T. Grube, H. Heinrichs, P. Stenzel, D. Stolten, Kosteneffiziente und klimagerechte Transformations-strategien für das deutsche Energiesystem bis zum Jahr 2050. (Kurzfassung), Forschungszentrum Jülich GmbH, Jülich, 2019. https://www.fz-juelich.de/iek/iek-3/DE/_Documents/Downloads/transformationStrategies2050_studySummary_2019-10-31.pdf.pdf (accessed November 29, 2019).

[7]   J. Noack, N. Roznyatovskaya, T. Herr, P. Fischer, The Chemistry of Redox-Flow Batteries, Angew. Chem. Int. Ed. Engl. 54 (2015) 9776–9809. https://doi.org/10.1002/anie.201410823.

[8]   H.S. Lim, A.M. Lackner, R.C. Knechtli, Zinc-Bromine Secondary Battery, J. Electrochem. Soc. 124 (1977) 1154–1157. https://doi.org/10.1149/1.2133517.

[9] D.J. Eustace, Bromine Complexation in Zinc-Bromine Circulating Batteries, J. Electrochem. Soc. 127 (1980) 528–532. https://doi.org/10.1149/1.2129706.

[10] L.W. Hruska, R.F. Savinell, Investigation of Factors Affecting Performance of the Iron-Redox Battery, J. Electrochem. Soc. 128 (1981) 18–25. https://doi.org/10.1149/1.2127366.

[11] B.S. Jayathilake, E.J. Plichta, M.A. Hendrickson, S.R. Narayanan, Improvements to the Coulombic Efficiency of the Iron Electrode for an All-Iron Redox-Flow Battery, J. Electrochem. Soc. 165 (2018) A1630–A1638. https://doi.org/10.1149/2.0451809jes.

[12] L. Sanz, D. Lloyd, E. Magdalena, J. Palma, K. Kontturi, Description and performance of a novel aqueous all-copper redox flow battery, J. Power Sources. 268 (2014) 121–128. https://doi.org/10.1016/j.jpowsour.2014.06.008.

[13] A. Hazza, D. Pletcher, R. Wills, A novel flow battery: A lead acid battery based on an electrolyte with soluble lead(II), Phys. Chem. Chem. Phys. 6 (2004) 1773–1778. https://doi.org/10.1039/B401115E.

[14] M.C. Tucker, V. Srinivasan, P.N. Ross, A.Z. Weber, Performance and cycling of the iron-ion/hydrogen redox flow cell with various catholyte salts, J Appl Electrochem. 43 (2013) 637–644. https://doi.org/10.1007/s10800-013-0553-2.

[15] M.C. Tucker, K.T. Cho, A.Z. Weber, Optimization of the iron-ion/hydrogen redox flow cell with iron chloride catholyte salt, J. Power Sources. 245 (2014) 691–697. https://doi.org/10.1016/j.jpowsour.2013.07.029.

[16] N.H. Hagedorn, NASA Redox Storage System Development Project - Final Report, NASA, 1984. https://ntrs.nasa.gov/archive/nasa/casi.ntrs.nasa.gov/19850004157.pdf.

[17] Y.-R. Dong, H. Kaku, K. Hanafusa, K. Moriuchi, T. Shigematsu, A Novel Titanium/Manganese Redox Flow Battery, ECS Trans. 69 (2015) 59–67. https://doi.org/10.1149/06918.0059ecst.

[18] H. Kaku, Y.-R. Dong, K. Hanafusa, K. Moriuchi, T. Shigematsu, Effect of Ti(IV) Ion on Mn(III) Stability in Ti/Mn Electrolyte for Redox Flow Battery, ECS Trans. 72 (2016) 1–9. https://doi.org/10.1149/07210.0001ecst.

[19] H. Kaku, Y. Kawagoe, Y.-R. Dong, R. Tatsumi, K. Moriuchi, T. Shigematsu, Enhanced Performance of Ti/Mn Redox Flow Battery, ECS Trans. 77 (2017) 173–183. https://doi.org/10.1149/07711.0173ecst.

[20] Y.-R. Dong, Y. Kawagoe, K. Itou, H. Kaku, K. Hanafusa, K. Moriuchi, T. Shigematsu, Improved Performance of Ti/Mn Redox Flow Battery by Thermally Treated Carbon Paper Electrodes, ECS Trans. 75 (2017) 27–35. https://doi.org/10.1149/07518.0027ecst.

[21] B. Huskinson, M.P. Marshak, C. Suh, S. Er, M.R. Gerhardt, C.J. Galvin, X. Chen, A. Aspuru-Guzik, R.G. Gordon, M.J. Aziz, A metal-free organic–inorganic aqueous flow battery, Nature. 505 (2014) 195–198. https://doi.org/10.1038/nature12909.

[22] B. Yang, L. Hoober-Burkhardt, F. Wang, G.K.S. Prakash, S.R. Narayanan, An Inexpensive Aqueous Flow Battery for Large-Scale Electrical Energy Storage Based on Water-Soluble Organic Redox Couples, J. Electrochem. Soc. 161 (2014) A1371–A1380. https://doi.org/10.1149/2.1001409jes.

[23] K. Lin, Q. Chen, M.R. Gerhardt, L. Tong, S.B. Kim, L. Eisenach, A.W. Valle, D. Hardee, R.G. Gordon, M.J. Aziz, M.P. Marshak, Alkaline quinone flow battery, Science. 349 (2015) 1529–1532. https://doi.org/10.1126/science.aab3033.

[24] M. Skyllas-Kazacos, M. Rychcik, R.G. Robins, A.G. Fane, M.A. Green, New All-Vanadium Redox Flow Cell, J. Electrochem. Soc. 133 (1986) 1057–1058. https://doi.org/10.1149/1.2108706.

[25] E. Sum, M. Rychcik, M. Skyllas-kazacos, Investigation of the V(V)/V(IV) system for use in the positive half-cell of a redox battery, J. Power Sources. 16 (1985) 85–95. https://doi.org/10.1016/0378-7753(85)80082-3.

[26] E. Sum, M. Skyllas-Kazacos, A study of the V(II)/V(III) redox couple for redox flow cell applications, J. Power Sources. 15 (1985) 179–190. https://doi.org/10.1016/0378-7753(85)80071-9.

[27] M. Skyllas-Kazacos, F. Grossmith, Efficient Vanadium Redox Flow Cell, J. Electrochem. Soc. 134 (1987) 2950–2953. https://doi.org/10.1149/1.2100321.

[28] B. Sun, M. Skyllas-Kazacos, Modification of graphite electrode materials for vanadium redox flow battery application—I. Thermal treatment, Electrochim. Acta. 37 (1992) 1253–1260. https://doi.org/10.1016/0013-4686(92)85064-R.

[29] B. Sun, M. Skyllas-Kazacos, Chemical modification of graphite electrode materials for vanadium redox flow battery application—part II. Acid treatments, Electrochim. Acta. 37 (1992) 2459–2465. https://doi.org/10.1016/0013-4686(92)87084-D.

[30] M. Kazacos, M. Cheng, M. Skyllas-Kazacos, Vanadium redox cell electrolyte optimization studies, J. Appl. Electrochem. 20 (1990) 463–467. https://doi.org/10.1007/BF01076057.

[31] C. Menictas, M. Cheng, M. Skyllas-Kazacos, Evaluation of an NH_4VO_3-derived electrolyte for the vanadium-redox flow battery, J. Power Sources. 45 (1993) 43–54. https://doi.org/10.1016/0378-7753(93)80007-C.

[32] M. Skyllas-Kazacos, C. Menictas, M. Kazacos, Thermal Stability of Concentrated V(V) Electrolytes in the Vanadium Redox Cell, J. Electrochem. Soc. 143 (1996) L86–L88. https://doi.org/10.1149/1.1836609.

[33] M. Duduta, B. Ho, V.C. Wood, P. Limthongkul, V.E. Brunini, W.C. Carter, Y.-M. Chiang, Semi-Solid Lithium Rechargeable Flow Battery, Adv. Energy Mater. 1 (2011) 511–516. https://doi.org/10.1002/aenm.201100152.

[34] J. Winsberg, T. Hagemann, T. Janoschka, M.D. Hager, U.S. Schubert, Redox-Flow Batteries: From Metals to Organic Redox-Active Materials., Angew Chem Int Ed Engl. 56 (2017) 686–711. https://doi.org/10.1002/anie.201604925.

[35] A. Khataee, K. Wedege, E. Dražević, A. Bentien, Differential pH as a method for increasing cell potential in organic aqueous flow batteries, J. Mater. Chem. A. 5 (2017) 21875–21882. https://doi.org/10.1039/C7TA04975G.

[36] H. Chen, G. Cong, Y.-C. Lu, Recent progress in organic redox flow batteries: Active materials, electrolytes and membranes, J. Energy Chem. 27 (2018) 1304–1325. https://doi.org/10.1016/j.jechem.2018.02.009.

[37] P. Leung, A.A. Shah, L. Sanz, C. Flox, J.R. Morante, Q. Xu, M.R. Mohamed, C. Ponce de León, F.C. Walsh, Recent developments in organic redox flow batteries: A critical review, J. Power Sources. 360 (2017) 243–283. https://doi.org/10.1016/j.jpowsour.2017.05.057.

[38] N. Kausar, A. Mousa, M. Skyllas-Kazacos, The Effect of Additives on the High-Temperature Stability of the Vanadium Redox Flow Battery Positive Electrolytes, ChemElectroChem. 3 (2016) 276–282. https://doi.org/10.1002/celc.201500453.

[39] K. Wang, Y. Zhang, L. Liu, J. Xi, Z. Wu, X. Qiu, Broad temperature adaptability of vanadium redox flow battery-Part 3: The effects of total vanadium concentration and sulfuric acid concentration, Electrochim. Acta. 259 (2018) 11–19. https://doi.org/10.1016/j.electacta.2017.10.148.

[40] L. Li, S. Kim, W. Wang, M. Vijayakumar, Z. Nie, B. Chen, J. Zhang, G. Xia, J. Hu, G. Graff, J. Liu, Z. Yang, A Stable Vanadium Redox-Flow Battery with High Energy Density for Large-Scale Energy Storage, Adv. Energy Mater. 1 (2011) 394–400. https://doi.org/10.1002/aenm.201100008.

[41] A.A. Shah, M.J. Watt-Smith, F.C. Walsh, A dynamic performance model for redox-flow batteries involving soluble species, Electrochim. Acta. 53 (2008) 8087–8100. https://doi.org/10.1016/j.electacta.2008.05.067.

[42] M. Pavelka, F. Wandschneider, P. Mazur, Thermodynamic derivation of open circuit voltage in vanadium redox flow batteries, Journal of Power Sources. 293 (2015) 400–408. https://doi.org/10.1016/j.jpowsour.2015.05.049.

[43] S. Corcuera, M. Skyllas-Kazacos, State-of-Charge Monitoring and Electrolyte Rebalancing Methods for the Vanadium Redox Flow Battery, Eur. Chem. Bull. 1 (2012) 511–519. https://doi.org/10.17628/ecb.2012.1.511-519.

[44] X. Li, J. Xiong, A. Tang, Y. Qin, J. Liu, C. Yan, Investigation of the use of electrolyte viscosity for online state-of-charge monitoring design in vanadium redox flow battery, Applied Energy. 211 (2018) 1050–1059. https://doi.org/10.1016/j.apenergy.2017.12.009.

[45] U.S. Department of Energy, Grid Energy Storage, (2013). https://www.energy.gov /sites/prod/files/2014/09/f18/Grid%20Energy%20Storage%20December%202013.pdf (accessed June 24, 2018).

[46] Breaking News Of Vanadium Pentoxide - Mining Companies In Canada, (2019). https://www.vanadiumprice.com/ (accessed November 27, 2019).

[47] Euro Dollar Exchange Rate (EUR USD) - Historical Chart, (n.d.). https://www.macrotrends.net/2548/euro-dollar-exchange-rate-historical-chart (accessed November 27, 2019).

[48] Mineral commodity summaries 2019, U.S. Geological Survey, 2019. https://prd-wret.s3-us-west-2.amazonaws.com/assets/palladium/production/atoms/files/mcs2019_all.pdf (accessed December 3, 2019).

[49] Q. Xu, T.S. Zhao, Fundamental models for flow batteries, Prog. Energ. Combust. 49 (2015) 40–58. https://doi.org/10.1016/j.pecs.2015.02.001.

[50] B. Xiong, J. Zhao, Z. Wei, M. Skyllas-Kazacos, Extended Kalman filter method for state of charge estimation of vanadium redox flow battery using thermal-dependent electrical model, J. Power Sources. 262 (2014) 50–61. https://doi.org/10.1016/j.jpowsour.2014.03.110.

[51] Z. Wei, K.J. Tseng, N. Wai, T.M. Lim, M. Skyllas-Kazacos, Adaptive estimation of state of charge and capacity with online identified battery model for vanadium redox flow battery, J. Power Sources. 332 (2016) 389–398. https://doi.org/10.1016/j.jpowsour.2016.09.123.

[52] Z. Wei, T.M. Lim, M. Skyllas-Kazacos, N. Wai, K.J. Tseng, Online state of charge and model parameter co-estimation based on a novel multi-timescale estimator for vanadium redox flow battery, Appl. Energy. 172 (2016) 169–179. https://doi.org/10.1016/j.apenergy.2016.03.103.

[53] Z. Wei, S. Meng, K.J. Tseng, T.M. Lim, B.H. Soong, M. Skyllas-Kazacos, An adaptive model for vanadium redox flow battery and its application for online peak power estimation, J. Power Sources. 344 (2017) 195–207. https://doi.org/10.1016/j.jpowsour.2017.01.102.

[54] Z. Wei, A. Bhattarai, C. Zou, S. Meng, T.M. Lim, M. Skyllas-Kazacos, Real-time monitoring of capacity loss for vanadium redox flow battery, J. Power Sources. 390 (2018) 261–269. https://doi.org/10.1016/j.jpowsour.2018.04.063.

[55] F. Xing, H. Zhang, X. Ma, Shunt current loss of the vanadium redox flow battery, J. Power Sources. 196 (2011) 10753–10757. https://doi.org/10.1016/j.jpowsour.2011.08.033.

[56] F.T. Wandschneider, S. Röhm, P. Fischer, K. Pinkwart, J. Tübke, H. Nirschl, A multi-stack simulation of shunt currents in vanadium redox flow batteries, J. Power Sources. 261 (2014) 64–74. https://doi.org/10.1016/j.jpowsour.2014.03.054.

[57] H. Fink, M. Remy, Shunt currents in vanadium flow batteries: Measurement, modelling and implications for efficiency, J. Power Sources. 284 (2015) 547–553. https://doi.org/10.1016/j.jpowsour.2015.03.057.

[58] M. Skyllas-Kazacos, J. McCann, Y. Li, J. Bao, A. Tang, The Mechanism and Modelling of Shunt Current in the Vanadium Redox Flow Battery, ChemistrySelect. 1 (2016) 2249–2256. https://doi.org/10.1002/slct.201600432.

[59] Y.-S. Chen, S.-Y. Ho, H.-W. Chou, H.-J. Wei, Modeling the effect of shunt current on the charge transfer efficiency of an all-vanadium redox flow battery, J. Power Sources. 390 (2018) 168–175. https://doi.org/10.1016/j.jpowsour.2018.04.042.

[60] A. Tang, J. Bao, M. Skyllas-Kazacos, Dynamic modelling of the effects of ion diffusion and side reactions on the capacity loss for vanadium redox flow battery, J. Power Sources. 196 (2011) 10737–10747. https://doi.org/10.1016/j.jpowsour.2011.09.003.

[61] D. You, H. Zhang, C. Sun, X. Ma, Simulation of the self-discharge process in vanadium redox flow battery, J. Power Sources. 196 (2011) 1578–1585. https://doi.org/10.1016/j.jpowsour.2010.08.036.

[62] M. Skyllas-Kazacos, L. Goh, Modeling of vanadium ion diffusion across the ion exchange membrane in the vanadium redox battery, J. Membrane Sci. 399–400 (2012) 43–48. https://doi.org/10.1016/j.memsci.2012.01.024.

[63] R. Badrinarayanan, J. Zhao, K.J. Tseng, M. Skyllas-Kazacos, Extended dynamic model for ion diffusion in all-vanadium redox flow battery including the effects of temperature and bulk electrolyte transfer, J. Power Sources. 270 (2014) 576–586. https://doi.org/10.1016/j.jpowsour.2014.07.128.

[64] K. Schafner, M. Becker, T. Turek, Capacity balancing for vanadium redox flow batteries through electrolyte overflow, J Appl Electrochem. 48 (2018) 639–649. https://doi.org/10.1007/s10800-018-1187-1.

[65] L.J. Ontiveros, P.E. Mercado, Modeling of a Vanadium Redox Flow Battery for power system dynamic studies, Int. J. Hydrogen Energ. 39 (2014) 8720–8727. https://doi.org/10.1016/j.ijhydene.2013.12.042.

[66] A. Tang, J. Bao, M. Skyllas-Kazacos, Studies on pressure losses and flow rate optimization in vanadium redox flow battery, J. Power Sources. 248 (2014) 154–162. https://doi.org/10.1016/j.jpowsour.2013.09.071.

[67] S. König, M.R. Suriyah, T. Leibfried, Model based examination on influence of stack series connection and pipe diameters on efficiency of vanadium redox flow batteries under

consideration of shunt currents, J. Power Sources. 281 (2015) 272–284. https://doi.org/10.1016/j.jpowsour.2015.01.119.

[68] Q. Ye, J. Hu, P. Cheng, Z. Ma, Design trade-offs among shunt current, pumping loss and compactness in the piping system of a multi-stack vanadium flow battery, J. Power Sources. 296 (2015) 352–364. https://doi.org/10.1016/j.jpowsour.2015.06.138.

[69] Y. Zhang, J. Zhao, P. Wang, M. Skyllas-Kazacos, B. Xiong, R. Badrinarayanan, A comprehensive equivalent circuit model of all-vanadium redox flow battery for power system analysis, J. Power Sources. 290 (2015) 14–24. https://doi.org/10.1016/j.jpowsour.2015.04.169.

[70] J. Fu, T. Wang, X. Wang, J. Sun, M. Zheng, Dynamic Flow Rate Control for Vanadium Redox Flow Batteries, Energy Procedia. 105 (2017) 4482–4491. https://doi.org/10.1016/j.egypro.2017.03.952.

[71] Y. Li, X. Zhang, J. Bao, M. Skyllas-Kazacos, Control of electrolyte flow rate for the vanadium redox flow battery by gain scheduling, J. Energy Storage. 14 (2017) 125–133. https://doi.org/10.1016/j.est.2017.10.005.

[72] Y. Li, X. Zhang, J. Bao, M. Skyllas-Kazacos, Studies on optimal charging conditions for vanadium redox flow batteries, J. Energy Storage. 11 (2017) 191–199. https://doi.org/10.1016/j.est.2017.02.008.

[73] T. Wang, J. Fu, M. Zheng, Z. Yu, Dynamic control strategy for the electrolyte flow rate of vanadium redox flow batteries, Appl. Energy. (2018) 613–623. https://doi.org/10.1016/j.apenergy.2017.07.065.

[74] A. Tang, S. Ting, J. Bao, M. Skyllas-Kazacos, Thermal modelling and simulation of the all-vanadium redox flow battery, J. Power Sources. 203 (2012) 165–176. https://doi.org/10.1016/j.jpowsour.2011.11.079.

[75] B. Xiong, J. Zhao, K.J. Tseng, M. Skyllas-Kazacos, T.M. Lim, Y. Zhang, Thermal hydraulic behavior and efficiency analysis of an all-vanadium redox flow battery, J. Power Sources. 242 (2013) 314–324. https://doi.org/10.1016/j.jpowsour.2013.05.092.

[76] Z. Wei, J. Zhao, B. Xiong, Dynamic electro-thermal modeling of all-vanadium redox flow battery with forced cooling strategies, Appl. Energy. 135 (2014) 1–10. https://doi.org/10.1016/j.apenergy.2014.08.062.

[77] Y. Yan, Y. Li, M. Skyllas-Kazacos, J. Bao, Modelling and simulation of thermal behaviour of vanadium redox flow battery, J. Power Sources. 322 (2016) 116–128. https://doi.org/10.1016/j.jpowsour.2016.05.011.

[78] Y. Yan, M. Skyllas-Kazacos, J. Bao, Effects of battery design, environmental temperature and electrolyte flowrate on thermal behaviour of a vanadium redox flow battery in different applications, J. Energy Storage. 11 (2017) 104–118. https://doi.org/10.1016/j.est.2017.01.007.

[79] A.K. Sharma, C.Y. Ling, E. Birgersson, M. Vynnycky, M. Han, Verified reduction of dimensionality for an all-vanadium redox flow battery model, J. Power Sources. 279 (2015) 345–350. https://doi.org/10.1016/j.jpowsour.2015.01.019.

[80] M. Vynnycky, Analysis of a model for the operation of a vanadium redox battery, Energy. 36 (2011) 2242–2256. https://doi.org/10.1016/j.energy.2010.03.060.

[81] C.L. Chen, H.K. Yeoh, M.H. Chakrabarti, An enhancement to Vynnycky's model for the all-vanadium redox flow battery, Electrochim. Acta. 120 (2014) 167–179. https://doi.org/10.1016/j.electacta.2013.12.074.

[82] Y. Lei, B.W. Zhang, B.F. Bai, T.S. Zhao, A transient electrochemical model incorporating the Donnan effect for all-vanadium redox flow batteries, J. Power Sources. 299 (2015) 202–211. https://doi.org/10.1016/j.jpowsour.2015.08.100.

[83] Y. Lei, B.W. Zhang, Z.H. Zhang, B.F. Bai, T.S. Zhao, An improved model of ion selective adsorption in membrane and its application in vanadium redox flow batteries, Appl. Energy. 215 (2018) 591–601. https://doi.org/10.1016/j.apenergy.2018.02.042.

[84] C.-N. Sun, F.M. Delnick, D.S. Aaron, A.B. Papandrew, M.M. Mench, T.A. Zawodzinski, Resolving Losses at the Negative Electrode in All-Vanadium Redox Flow Batteries Using Electrochemical Impedance Spectroscopy, J. Electrochem. Soc. 161 (2014) A981–A988. https://doi.org/10.1149/2.045406jes.

[85] A.M. Pezeshki, R.L. Sacci, F.M. Delnick, D.S. Aaron, M.M. Mench, Elucidating effects of cell architecture, electrode material, and solution composition on overpotentials in redox flow batteries, Electrochim. Acta. 229 (2017) 261–270. https://doi.org/10.1016/j.electacta.2017.01.056.

[86] M. Zago, A. Casalegno, Physically-based impedance modeling of the negative electrode in All-Vanadium Redox Flow Batteries: insight into mass transport issues, Electrochim. Acta. 248 (2017) 505–517. https://doi.org/10.1016/j.electacta.2017.07.166.

[87] Y. Li, M. Skyllas-Kazacos, J. Bao, A dynamic plug flow reactor model for a vanadium redox flow battery cell, J. Power Sources. 311 (2016) 57–67. https://doi.org/10.1016/j.jpowsour.2016.02.018.

[88] S. König, M.R. Suriyah, T. Leibfried, A plug flow reactor model of a vanadium redox flow battery considering the conductive current collectors, J. Power Sources. 360 (2017) 221–231. https://doi.org/10.1016/j.jpowsour.2017.05.085.

[89] H. Al-Fetlawi, A.A. Shah, F.C. Walsh, Non-isothermal modelling of the all-vanadium redox flow battery, Electrochim. Acta. 55 (2009) 78–89. https://doi.org/10.1016/j.electacta.2009.08.009.

[90] A.A. Shah, H. Al-Fetlawi, F.C. Walsh, Dynamic modelling of hydrogen evolution effects in the all-vanadium redox flow battery, Electrochim. Acta. 55 (2010) 1125–1139. https://doi.org/10.1016/j.electacta.2009.10.022.

[91] H. Al-Fetlawi, A.A. Shah, F.C. Walsh, Modelling the effects of oxygen evolution in the all-vanadium redox flow battery, Electrochim. Acta. 55 (2010) 3192–3205. https://doi.org/10.1016/j.electacta.2009.12.085.

[92] M.J. Watt-Smith, P. Ridley, R.G.A. Wills, A.A. Shah, F.C. Walsh, The importance of key operational variables and electrolyte monitoring to the performance of an all vanadium redox flow battery, J. Chem. Technol. Biotechnol. 88 (2013) 126–138. https://doi.org/10.1002/jctb.3870.

[93] D. You, H. Zhang, J. Chen, A simple model for the vanadium redox battery, Electrochim. Acta. 54 (2009) 6827–6836. https://doi.org/10.1016/j.electacta.2009.06.086.

[94] D. Schmal, J. Van Erkel, P.J. Van Duin, Mass transfer at carbon fibre electrodes, J. Appl. Electrochem. 16 (1986) 422–430. https://doi.org/10.1007/BF01008853.

[95] K.W. Knehr, E. Agar, C.R. Dennison, A.R. Kalidindi, E.C. Kumbur, A Transient Vanadium Flow Battery Model Incorporating Vanadium Crossover and Water Transport through the Membrane, J. Electrochem. Soc. 159 (2012) A1446–A1459. https://doi.org/10.1149/2.017209jes.

[96] E. Agar, K.W. Knehr, D. Chen, M.A. Hickner, E.C. Kumbur, Species transport mechanisms governing capacity loss in vanadium flow batteries: Comparing Nafion® and sulfonated Radel membranes, Electrochim. Acta. 98 (2013) 66–74. https://doi.org/10.1016/j.electacta.2013.03.030.

[97] K. Bromberger, J. Kaunert, T. Smolinka, A Model for All-Vanadium Redox Flow Batteries: Introducing Electrode-Compression Effects on Voltage Losses and Hydraulics, Energy Technol. 2 (2014) 64–76. https://doi.org/10.1002/ente.201300114.

[98] X.L. Zhou, T.S. Zhao, L. An, Y.K. Zeng, X.H. Yan, A vanadium redox flow battery model incorporating the effect of ion concentrations on ion mobility, Appl. Energy. 158 (2015) 157–166. https://doi.org/10.1016/j.apenergy.2015.08.028.

[99] S. Rudolph, U. Schröder, I.M. Bayanov, D. Hage, Measurement, simulation and in situ regeneration of energy efficiency in vanadium redox flow battery, J. Electroanalyt. Chem. 728 (2014) 72–80. https://doi.org/10.1016/j.jelechem.2014.05.033.

[100] Q. Xu, T.S. Zhao, C. Zhang, Effects of SOC-dependent electrolyte viscosity on performance of vanadium redox flow batteries, Appl. Energy. 130 (2014) 139–147. https://doi.org/10.1016/j.apenergy.2014.05.034.

[101] F.T. Wandschneider, D. Finke, S. Grosjean, P. Fischer, K. Pinkwart, J. Tübke, H. Nirschl, Model of a vanadium redox flow battery with an anion exchange membrane and a Larminie-correction, J. Power Sources. 272 (2014) 436–447. https://doi.org/10.1016/j.jpowsour.2014.08.082.

[102] A. Khazaeli, A. Vatani, N. Tahouni, M.H. Panjeshahi, Numerical investigation and thermodynamic analysis of the effect of electrolyte flow rate on performance of all

vanadium redox flow batteries, J. Power Sources. 293 (2015) 599–612. https://doi.org/10.1016/j.jpowsour.2015.05.100.

[103]   X.-G. Yang, Q. Ye, P. Cheng, T.S. Zhao, Effects of the electric field on ion crossover in vanadium redox flow batteries, Appl. Energy. 145 (2015) 306–319. https://doi.org/10.1016/j.apenergy.2015.02.038.

[104]   R.M. Darling, A.Z. Weber, M.C. Tucker, M.L. Perry, The Influence of Electric Field on Crossover in Redox-Flow Batteries, J. Electrochem. Soc. 163 (2016) A5014–A5022. https://doi.org/10.1149/2.0031601jes.

[105]   W.W. Yang, Y.L. He, Y.S. Li, Performance Modeling of a Vanadium Redox Flow Battery during Discharging, Electrochim. Acta. 155 (2015) 279–287. https://doi.org/10.1016/j.electacta.2014.12.138.

[106]   W.W. Yang, F.Y. Yan, Z.G. Qu, Y.L. He, Effect of various strategies of soc-dependent operating current on performance of a vanadium redox flow battery, Electrochim. Acta. 259 (2018) 772–782. https://doi.org/10.1016/j.electacta.2017.10.201.

[107]   G. Qiu, A.S. Joshi, C.R. Dennison, K.W. Knehr, E.C. Kumbur, Y. Sun, 3-D pore-scale resolved model for coupled species/charge/fluid transport in a vanadium redox flow battery, Electrochim. Acta. 64 (2012) 46–64. https://doi.org/10.1016/j.electacta.2011.12.065.

[108]   G. Qiu, C.R. Dennison, K.W. Knehr, E.C. Kumbur, Y. Sun, Pore-scale analysis of effects of electrode morphology and electrolyte flow conditions on performance of vanadium redox flow batteries, J. Power Sources. 219 (2012) 223–234. https://doi.org/10.1016/j.jpowsour.2012.07.042.

[109]   Q. Xu, T.S. Zhao, P.K. Leung, Numerical investigations of flow field designs for vanadium redox flow batteries, Appl. Energy. 105 (2013) 47–56. https://doi.org/10.1016/j.apenergy.2012.12.041.

[110]   C. Yin, Y. Gao, S. Guo, H. Tang, A coupled three dimensional model of vanadium redox flow battery for flow field designs, Energy. 74 (2014) 886–895. https://doi.org/10.1016/j.energy.2014.07.066.

[111]   S. Won, K. Oh, H. Ju, Numerical studies of carbon paper-based vanadium redox flow batteries, Electrochim. Acta. 201 (2016) 286–299. https://doi.org/10.1016/j.electacta.2015.11.091.

[112]   Q. Wang, Z.G. Qu, Z.Y. Jiang, W.W. Yang, Numerical study on vanadium redox flow battery performance with non-uniformly compressed electrode and serpentine flow field, Appl. Energy. 220 (2018) 106–116. https://doi.org/10.1016/j.apenergy.2018.03.058.

[113]   X. Ma, H. Zhang, F. Xing, A three-dimensional model for negative half cell of the vanadium redox flow battery, Electrochim. Acta. 58 (2011) 238–246. https://doi.org/10.1016/j.electacta.2011.09.042.

[114]   Q. Zheng, H. Zhang, F. Xing, X. Ma, X. Li, G. Ning, A three-dimensional model for thermal analysis in a vanadium flow battery, Appl. Energy. 113 (2014) 1675–1685. https://doi.org/10.1016/j.apenergy.2013.09.021.

[115]   K. Oh, S. Won, H. Ju, A comparative study of species migration and diffusion mechanisms in all-vanadium redox flow batteries, Electrochim. Acta. 181 (2015) 238–247. https://doi.org/10.1016/j.electacta.2015.03.012.

[116]   K. Oh, S. Won, H. Ju, Numerical study of the effects of carbon felt electrode compression in all-vanadium redox flow batteries, Electrochim. Acta. 181 (2015) 13–23. https://doi.org/10.1016/j.electacta.2015.02.212.

[117]   K. Oh, H. Yoo, J. Ko, S. Won, H. Ju, Three-dimensional, transient, nonisothermal model of all-vanadium redox flow batteries, Energy. 81 (2015) 3–14. https://doi.org/10.1016/j.energy.2014.05.020.

[118]   S. Won, K. Oh, H. Ju, Numerical analysis of vanadium crossover effects in all-vanadium redox flow batteries, Electrochim. Acta. 177 (2015) 310–320. https://doi.org/10.1016/j.electacta.2015.01.166.

[119]   S. König, M.R. Suriyah, T. Leibfried, Validating and improving a zero-dimensional stack voltage model of the Vanadium Redox Flow Battery, J. Power Sources. 378 (2018) 10–18. https://doi.org/10.1016/j.jpowsour.2017.12.014.

[120]   M. Pugach, M. Kondratenko, S. Briola, A. Bischi, Zero dimensional dynamic model of vanadium redox flow battery cell incorporating all modes of vanadium ions crossover, Appl. Energy. 226 (2018) 560–569. https://doi.org/10.1016/j.apenergy.2018.05.124.

[121]   Y.A. Gandomi, D.S. Aaron, T.A. Zawodzinski, M.M. Mench, In Situ Potential Distribution Measurement and Validated Model for All-Vanadium Redox Flow Battery, J. Electrochem. Soc. 163 (2016) A5188–A5201. https://doi.org/10.1149/2.0211601jes.

[122]   D. Aaron, C.-N. Sun, M. Bright, A.B. Papandrew, M.M. Mench, T.A. Zawodzinski, In Situ Kinetics Studies in All-Vanadium Redox Flow Batteries, ECS Electrochem. Lett. 2 (2013) A29–A31. https://doi.org/10.1149/2.001303eel.

[123]   Q. Liu, A. Turhan, T.A. Zawodzinski, M.M. Mench, In situ potential distribution measurement in an all-vanadium flow battery, Chem. Commun. 49 (2013) 6292–6294. https://doi.org/10.1039/C3CC42092B.

[124]   P. Mazúr, J. Mrlík, J. Beneš, J. Pocedič, J. Vrána, J. Dundálek, J. Kosek, Performance evaluation of thermally treated graphite felt electrodes for vanadium redox flow battery and their four-point single cell characterization, J. Power Sources. 380 (2018) 105–114. https://doi.org/10.1016/j.jpowsour.2018.01.079.

[125] L.A. Belfiore, Laminar Boundary Layer Mass Transfer around Solid Spheres, Gas Bubbles, and Other Submerged Objects, in: Transport Phenomena for Chemical Reactor Design, John Wiley & Sons, Ltd, 2003: pp. 275–360. https://doi.org/10.1002/0471471623.ch11.

[126] L.A. Belfiore, Dimensional Analysis of the Equations of Change for Fluid Dynamics Within the Mass Transfer Boundary Layer, in: Transport Phenomena for Chemical Reactor Design, John Wiley & Sons, Ltd, Hoboken, New Jersey, 2003: pp. 361–368. https://doi.org/10.1002/0471471623.ch12.

[127] R.H. Perry, D.W. Green, Mass transfer, in: Perry's Chemical Engineers' Handbook, 7th ed., Mcgraw-Hill Professional, New York, 1997: pp. 5–59.

[128] K. Kinoshita, S.C. Leach, Mass-Transfer Study of Carbon Felt, Flow-Through Electrode, J. Electrochem. Soc. 129 (1982) 1993–1997. https://doi.org/10.1149/1.2124338.

[129] I.M. Bayanov, R. Vanhaelst, The numerical simulation of vanadium RedOx flow batteries, J. Math. Chem. 49 (2011) 2013–2031. https://doi.org/10.1007/s10910-011-9872-x.

[130] Y. Wang, S.C. Cho, Analysis and Three-Dimensional Modeling of Vanadium Flow Batteries, J. Electrochem. Soc. 161 (2014) A1200–A1212. https://doi.org/10.1149/2.0061409jes.

[131] Y. Yu, Y. Zuo, C. Zuo, X. Liu, Z. Liu, A Hierarchical Multiscale Model for Microfluidic Fuel Cells with Porous Electrodes, Electrochim. Acta. 116 (2014) 237–243. https://doi.org/10.1016/j.electacta.2013.10.200.

[132] F.T. Wandschneider, M. Kuettinger, J. Noack, P. Fischer, K. Pinkwart, J. Tuebke, H. Nirschl, A coupled-physics model for the vanadium oxygen fuel cell, J. Power Sources. 259 (2014) 125–137. https://doi.org/10.1016/j.jpowsour.2014.02.087.

[133] V.K. Yu, D. Chen, Peak power prediction of a vanadium redox flow battery, J. Power Sources. 268 (2014) 261–268. https://doi.org/10.1016/j.jpowsour.2014.06.053.

[134] A. Crawford, E. Thomsen, D. Reed, D. Stephenson, V. Sprenkle, J. Liu, V. Viswanathan, Development and validation of chemistry agnostic flow battery cost performance model and application to nonaqueous electrolyte systems, Int. J. Energy Res. 40 (2016) 1611–1623. https://doi.org/10.1002/er.3526.

[135] M. Yue, Q. Zheng, F. Xing, H. Zhang, X. Li, X. Ma, Flow field design and optimization of high power density vanadium flow batteries: A novel trapezoid flow battery, AICHE J. 64 (2018) 782–795. https://doi.org/10.1002/aic.15959.

[136] S. Koenig, M.R. Suriyah, T. Leibfried, Validating and improving a zero-dimensional stack voltage model of the Vanadium Redox Flow Battery, J. Power Sources. 378 (2018) 10–18. https://doi.org/10.1016/j.jpowsour.2017.12.014.

[137]   Y. Kim, Y.Y. Choi, N. Yun, M. Yang, Y. Jeon, K.J. Kim, J.-I. Choi, Activity gradient carbon felt electrodes for vanadium redox flow batteries, J. Power Sources. 408 (2018) 128–135. https://doi.org/10.1016/j.jpowsour.2018.09.066.

[138]   B.W. Zhang, Y. Lei, B.F. Bai, A. Xu, T.S. Zhao, A two-dimensional mathematical model for vanadium redox flow battery stacks incorporating nonuniform electrolyte distribution in the flow frame, Appl. Therm. Eng. 151 (2019) 495–505. https://doi.org/10.1016/j.applthermaleng.2019.02.037.

[139]   R. Carta, S. Palmas, A.M. Polcaro, G. Tola, Behaviour of a carbon felt flow by electrodes Part I: Mass transfer characteristics, J. Appl. Electrochem. 21 (1991) 793–798. https://doi.org/10.1007/BF01402816.

[140]   Q. Xu, T.S. Zhao, Determination of the mass-transport properties of vanadium ions through the porous electrodes of vanadium redox flow batteries, Phys. Chem. Chem. Phys. 15 (2013) 10841–10848. https://doi.org/10.1039/C3CP51944A.

[141]   N. Vatistas, P.F. Marconi, M. Bartolozzi, Mass-transfer study of the carbon felt electrode, Electrochim. Acta. 36 (1991) 339–343. https://doi.org/10.1016/0013-4686(91)85259-A.

[142]   X. You, Q. Ye, P. Cheng, The Dependence of Mass Transfer Coefficient on the Electrolyte Velocity in Carbon Felt Electrodes: Determination and Validation, J. Electrochem. Soc. 164 (2017) E3386–E3394. https://doi.org/10.1149/2.0401711jes.

[143]   J.L. Barton, J.D. Milshtein, J.J. Hinricher, F.R. Brushett, Quantifying the impact of viscosity on mass-transfer coefficients in redox flow batteries, J. Power Sources. 399 (2018) 133–143. https://doi.org/10.1016/j.jpowsour.2018.07.046.

[144]   M.D.R. Kok, R. Jervis, T.G. Tranter, M.A. Sadeghi, D.J.L. Brett, P.R. Shearing, J.T. Gostick, Mass transfer in fibrous media with varying anisotropy for flow battery electrodes: Direct numerical simulations with 3D X-ray computed tomography, Chem. Eng. Sci. 196 (2019) 104–115. https://doi.org/10.1016/j.ces.2018.10.049.

[145]   F.W. Küster, A. Thiel, in: Rechentafeln Für Die Chemische Analytik, 105., Walter de Gruyter & Co. KG, Berlin; New York, 2002: p. 239.

[146]   G. Paasch, K. Micka, P. Gersdorf, Theory of the electrochemical impedance of macrohomogeneous porous electrodes, Electrochim. Acta. 38 (1993) 2653–2662. https://doi.org/10.1016/0013-4686(93)85083-B.

[147]   K. Schafner, M. Becker, T. Turek, Membrane resistance of different separator materials in a vanadium redox flow battery, J. Membrane Sci. (2019).

[148]   T. Yamamura, N. Watanabe, T. Yano, Y. Shiokawa, Electron-Transfer Kinetics of Np^{3+}/Np^{4+}, NpO_2^+/NpO_2^{2+}, V^{2+}/V^{3+}, and VO^{2+}/VO_2^+ at Carbon Electrodes, J. Electrochem. Soc. 152 (2005) A830. https://doi.org/10.1149/1.1870794.

[149] J.S. Lawton, S.M. Tiano, D.J. Donnelly, S.P. Flanagan, T.M. Arruda, The Effect of Sulfuric Acid Concentration on the Physical and Electrochemical Properties of Vanadyl Solutions, Batteries. 4 (2018) 40. https://doi.org/10.3390/batteries4030040.

[150] Z. Jiang, K. Klyukin, V. Alexandrov, Structure, hydrolysis, and diffusion of aqueous vanadium ions from Car-Parrinello molecular dynamics, J. Chem. Phys. 145 (2016) 114303. https://doi.org/10.1063/1.4962748.

[151] G. Oriji, Y. Katayama, T. Miura, Investigations on V(IV)/V(V) and V(II)/V(III) redox reactions by various electrochemical methods, J. Power Sources. 139 (2005) 321–324. https://doi.org/10.1016/j.jpowsour.2004.03.008.

[152] M. Gattrell, J. Park, B. MacDougall, J. Apte, S. McCarthy, C.W. Wu, Study of the Mechanism of the Vanadium 4+/5+ Redox Reaction in Acidic Solutions, J. Electrochem. Soc. 151 (2004) A123–A130. https://doi.org/10.1149/1.1630594.

Appendix

A.1 Detailed overview for measured and modeled potential probe signals

In chapter 6 a brief comparison of cell voltages and overpotentials measured experimentally during the recording of polarization curves and generated numerically by the mathematical model is given. In this section, the entire dataset for comparison of experimental and modelled data is presented. Figures A-1 to A-6 summarize the results for the cell setup with two untreated carbon felt electrodes (GFD4.6EA, SGL Carbon) with 9% compression rate, separated by a Nafion 117® (Chemours) membrane and two bipolar plates made from PPG86 (Eisenhuth). Figure A-7 to A-12 include the results from a cell setup with 42% carbon felt compression rate, but other parameters unchanged.

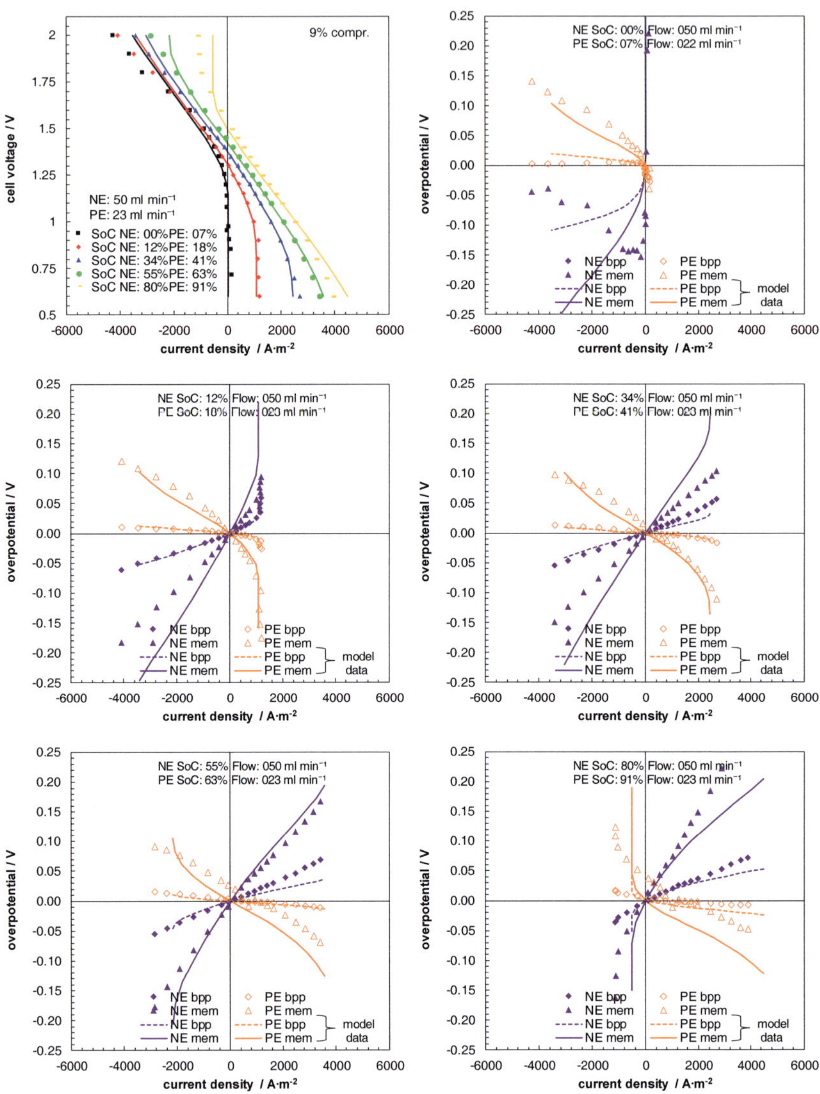

Figure A-1: Top left) Polarization curves of a cell setup with 9% carbon felt compression rate (GFD4.6E carbon felts, Nafion 117 membrane) for several states of charge. Others) Overpotentials measured at bipolar plate and membrane interface calculated by the difference between liquid and solid phase potential probe. Corresponding flow rate and SoC are shown in the figures for and PE, respectively. Experimental data represented as dots, whereas data from the model is indicated by lines.

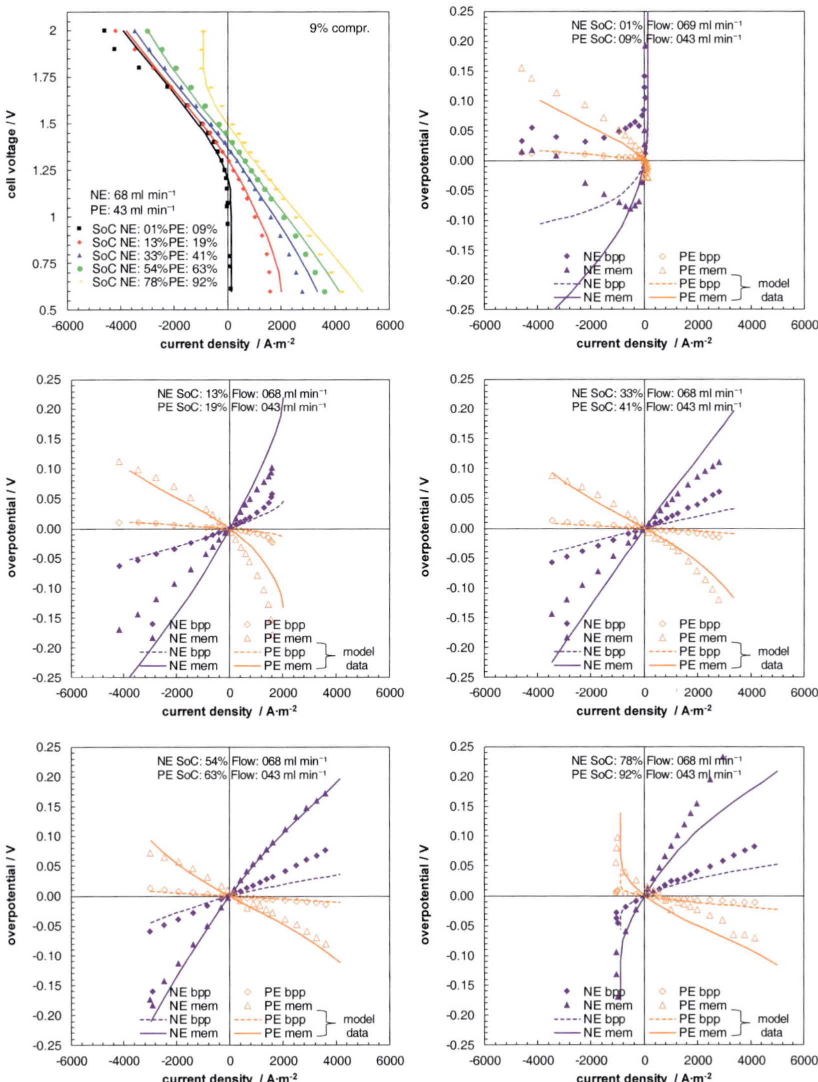

Figure A-2: Top left) Polarization curves of a cell setup with 9% carbon felt compression rate (GFD4.6E carbon felts, Nafion 117 membrane) for several states of charge. Others) Overpotentials measured at bipolar plate and membrane interface calculated by the difference between liquid and solid phase potential probe. Corresponding flow rate and SoC are shown in the figures for and PE, respectively. Experimental data represented as dots, whereas data from the model is indicated by lines.

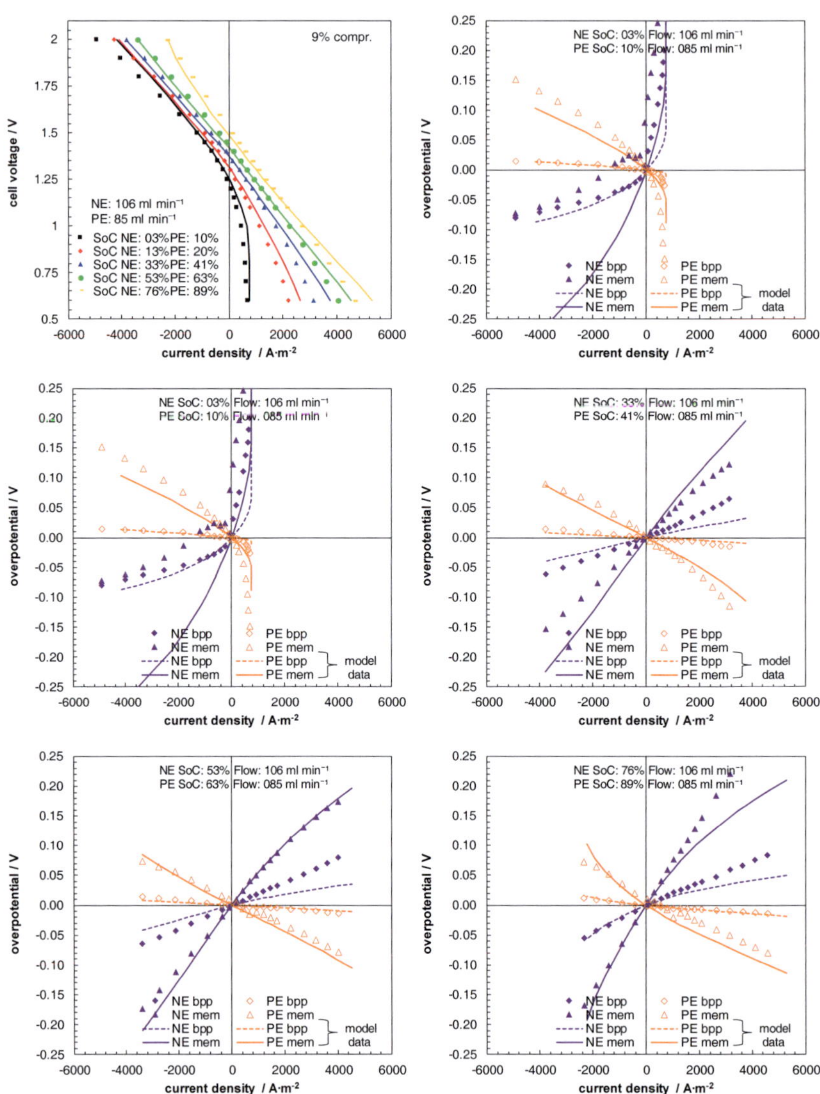

Figure A-3: Top left) Polarization curves of a cell setup with 9% carbon felt compression rate (GFD4.6E carbon felts, Nafion 117 membrane) for several states of charge. Others) Overpotentials measured at bipolar plate and membrane interface calculated by the difference between liquid and solid phase potential probe. Corresponding flow rate and SoC are shown in the figures for and PE, respectively. Experimental data represented as dots, whereas data from the model is indicated by lines.

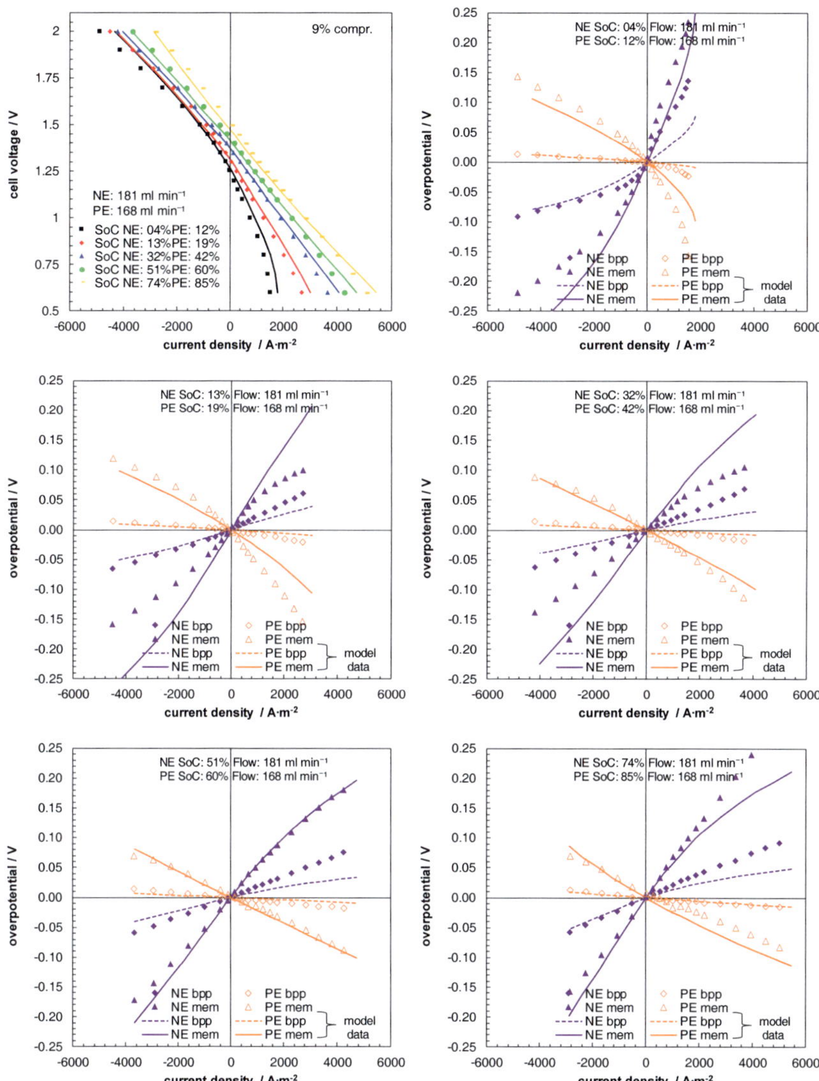

Figure A-4: Top left) Polarization curves of a cell setup with 9% carbon felt compression rate (GFD4.6E carbon felts, Nafion 117 membrane) for several states of charge. Others) Overpotentials measured at bipolar plate and membrane interface calculated by the difference between liquid and solid phase potential probe. Corresponding flow rate and SoC are shown in the figures for and PE, respectively. Experimental data represented as dots, whereas data from the model is indicated by lines.

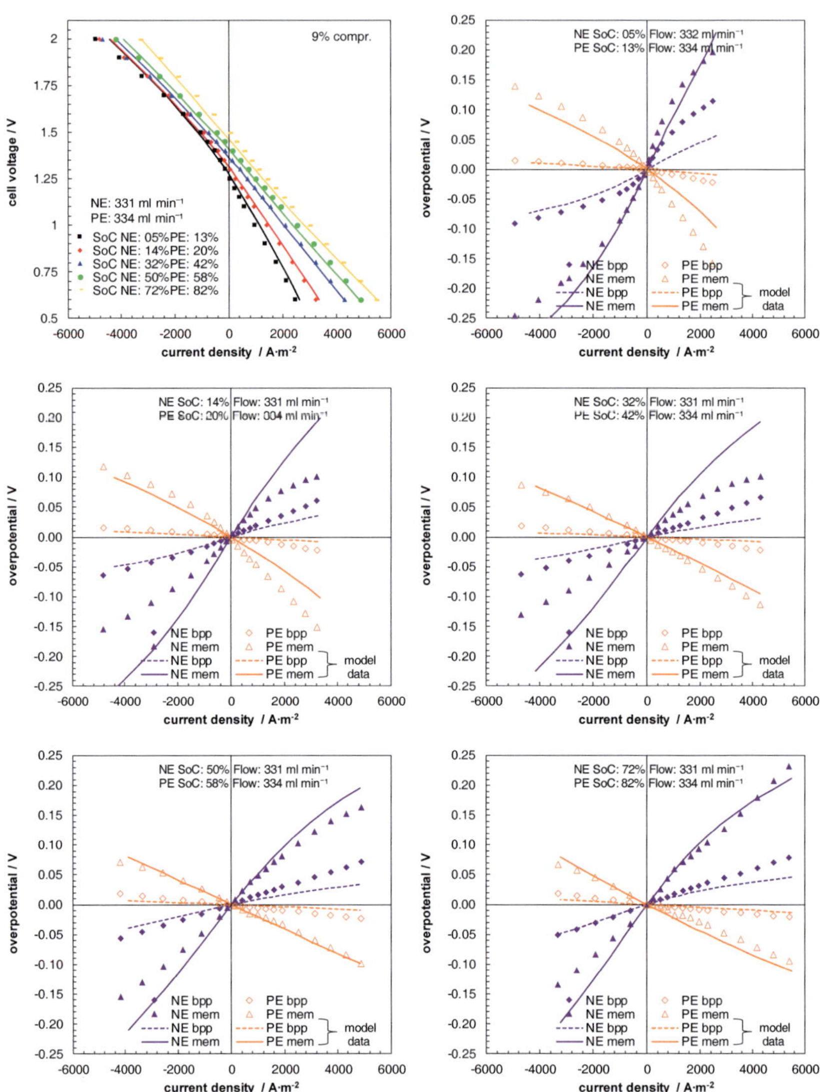

Figure A-5: Top left) Polarization curves of a cell setup with 9% carbon felt compression rate (GFD4.6E carbon felts, Nafion 117 membrane) for several states of charge. Others) Overpotentials measured at bipolar plate and membrane interface calculated by the difference between liquid and solid phase potential probe. Corresponding flow rate and SoC are shown in the figures for and PE, respectively. Experimental data represented as dots, whereas data from the model is indicated by lines.

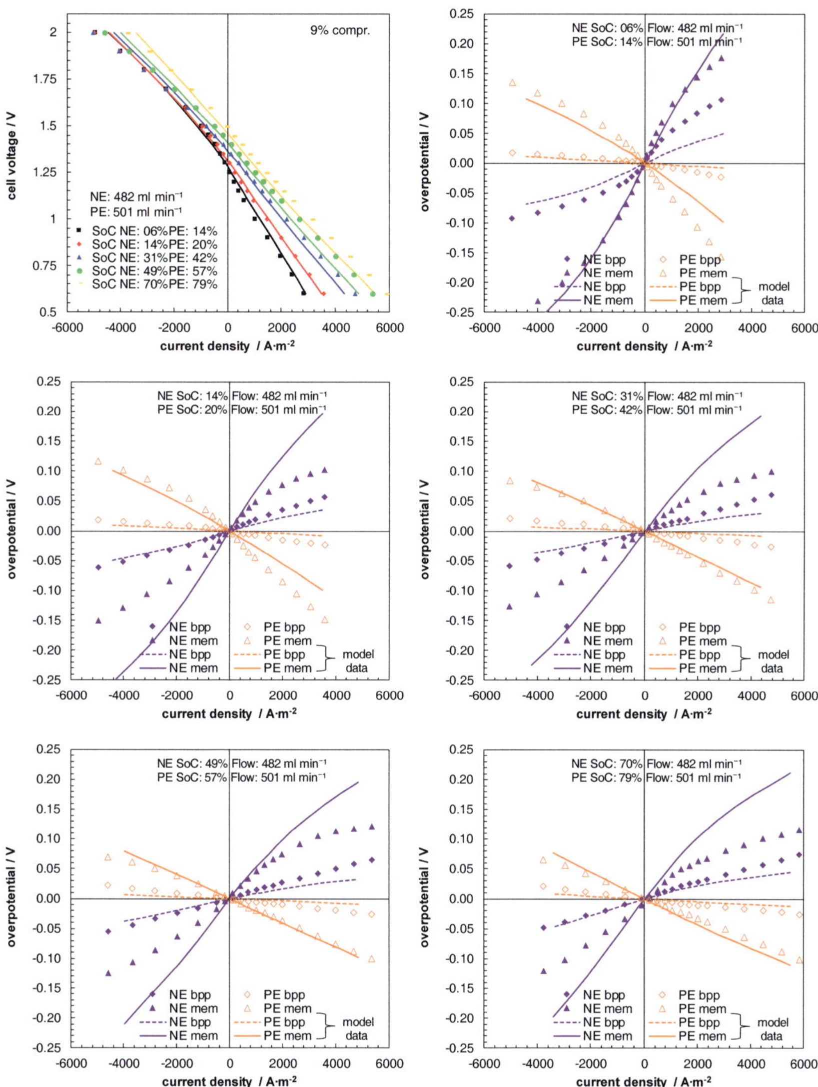

Figure A-6: Top left) Polarization curves of a cell setup with 9% carbon felt compression rate (GFD4.6E carbon felts, Nafion 117 membrane) for several states of charge. Others) Overpotentials measured at bipolar plate and membrane interface calculated by the difference between liquid and solid phase potential probe. Corresponding flow rate and SoC are shown in the figures for and PE, respectively. Experimental data represented as dots, whereas data from the model is indicated by lines.

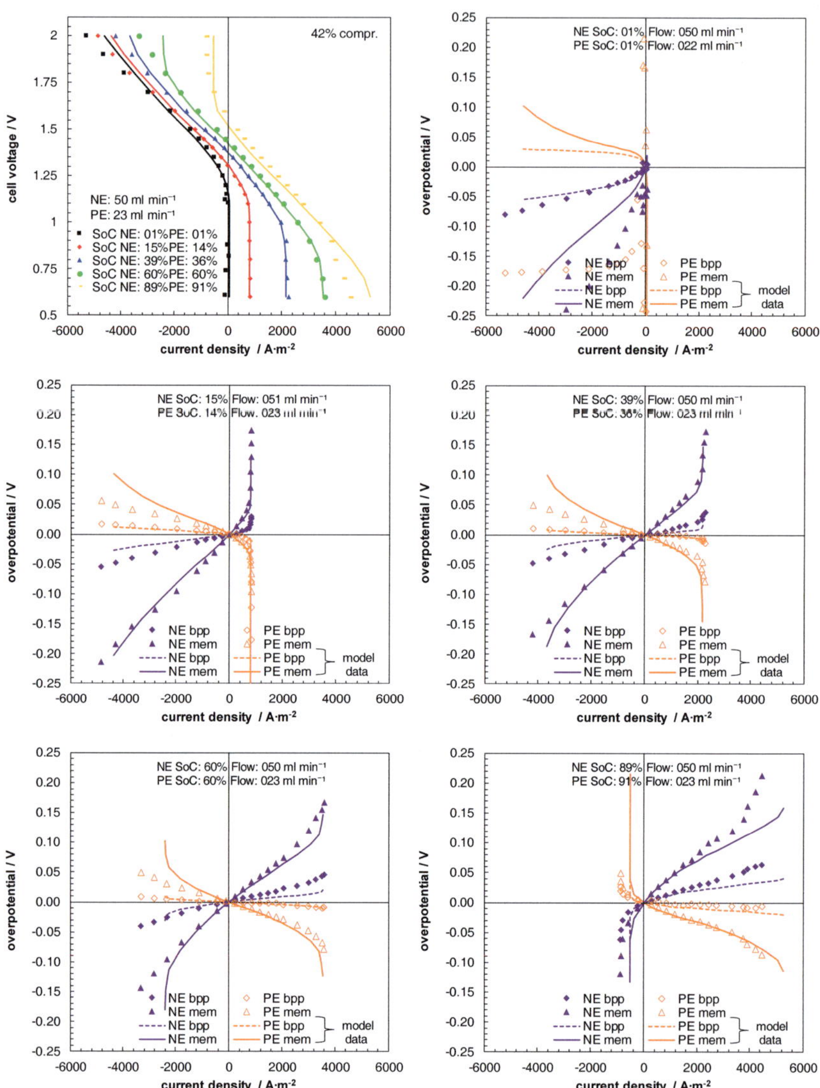

Figure A-7: Top left) Polarization curves of a cell setup with 42% carbon felt compression rate (GFD4.6E carbon felts, Nafion 117 membrane) for several states of charge. Others) Overpotentials measured at bipolar plate and membrane interface calculated by the difference between liquid and solid phase potential probe. Corresponding flow rate and SoC are shown in the figures for and PE, respectively. Experimental data represented as dots, whereas data from the model is indicated by lines.

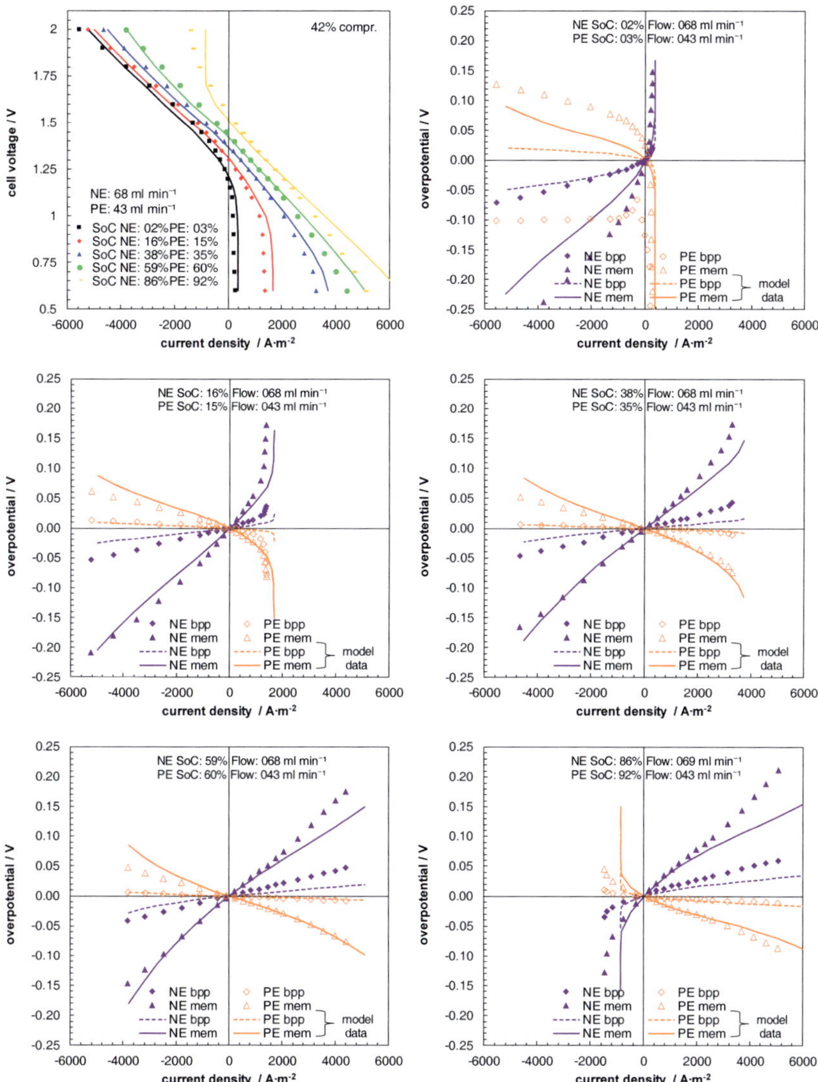

Figure A-8: Top left) Polarization curves of a cell setup with 42% carbon felt compression rate (GFD4.6E carbon felts, Nafion 117 membrane) for several states of charge. Others) Overpotentials measured at bipolar plate and membrane interface calculated by the difference between liquid and solid phase potential probe. Corresponding flow rate and SoC are shown in the figures for and PE, respectively. Experimental data represented as dots, whereas data from the model is indicated by lines.

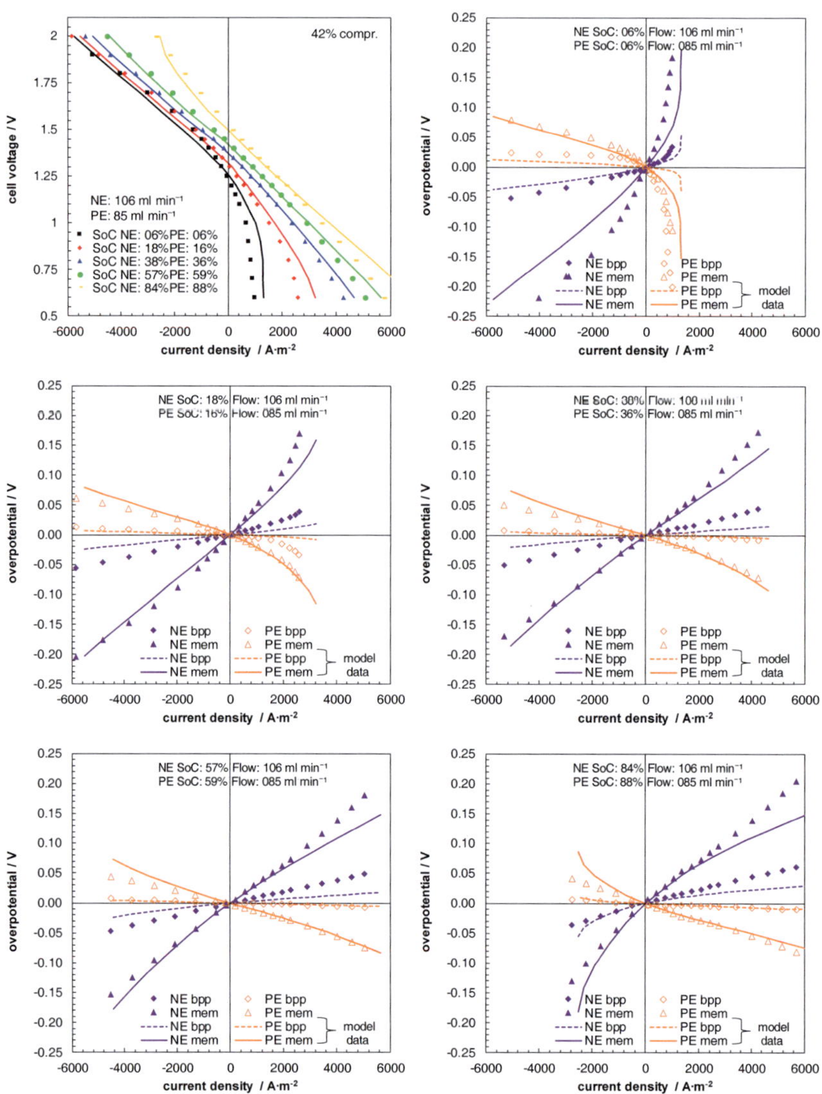

Figure A-9: Top left) Polarization curves of a cell setup with 42% carbon felt compression rate (GFD4.6E carbon felts, Nafion 117 membrane) for several states of charge. Others) Overpotentials measured at bipolar plate and membrane interface calculated by the difference between liquid and solid phase potential probe. Corresponding flow rate and SoC are shown in the figures for and PE, respectively. Experimental data represented as dots, whereas data from the model is indicated by lines.

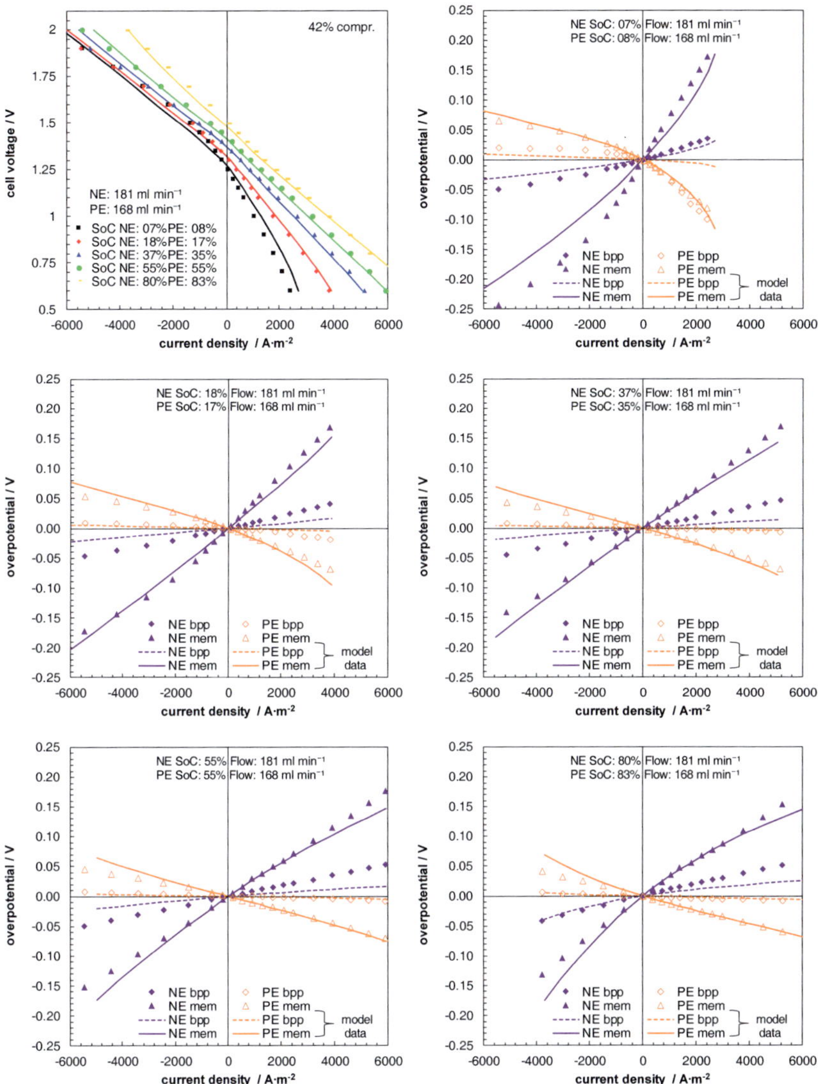

Figure A-10: Top left) Polarization curves of a cell setup with 42% carbon felt compression rate (GFD4.6E carbon felts, Nafion 117 membrane) for several states of charge. Others) Overpotentials measured at bipolar plate and membrane interface calculated by the difference between liquid and solid phase potential probe. Corresponding flow rate and SoC are shown in the figures for and PE, respectively. Experimental data represented as dots, whereas data from the model is indicated by lines.

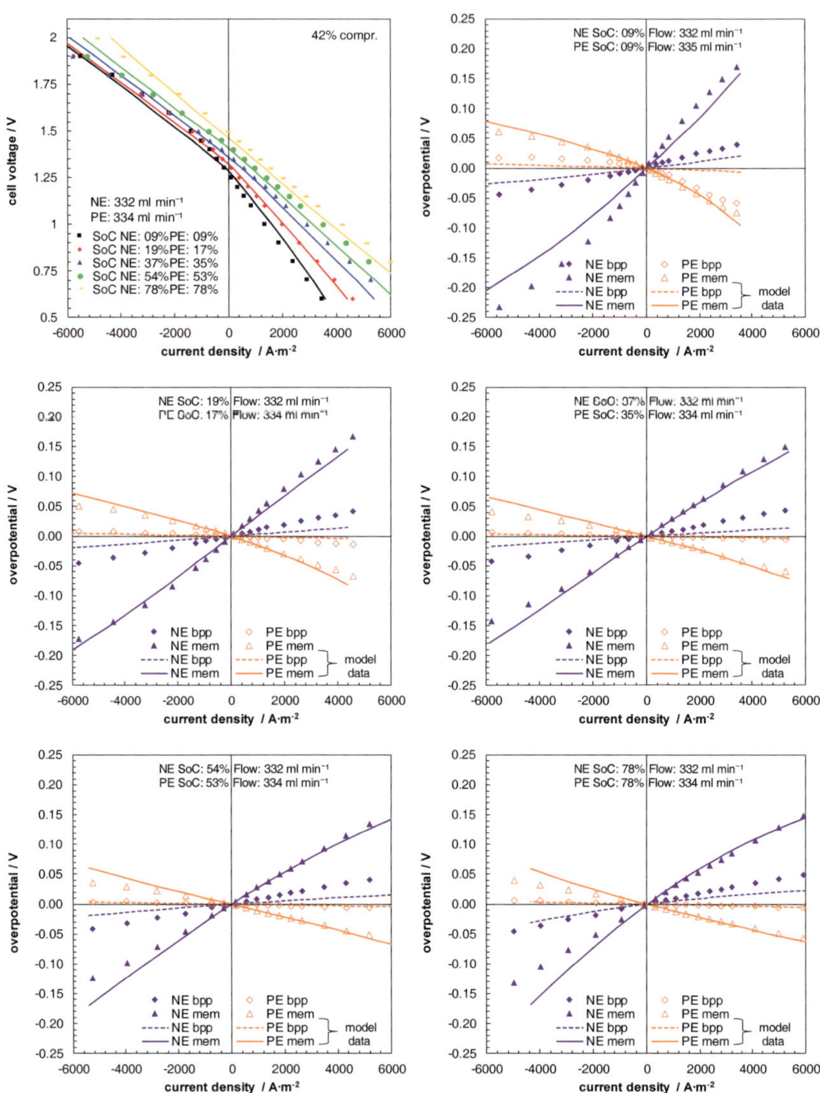

Figure A-11: Top left) Polarization curves of a cell setup with 42% carbon felt compression rate (GFD4.6E carbon felts, Nafion 117 membrane) for several states of charge. Others) Overpotentials measured at bipolar plate and membrane interface calculated by the difference between liquid and solid phase potential probe. Corresponding flow rate and SoC are shown in the figures for and PE, respectively. Experimental data represented as dots, whereas data from the model is indicated by lines.

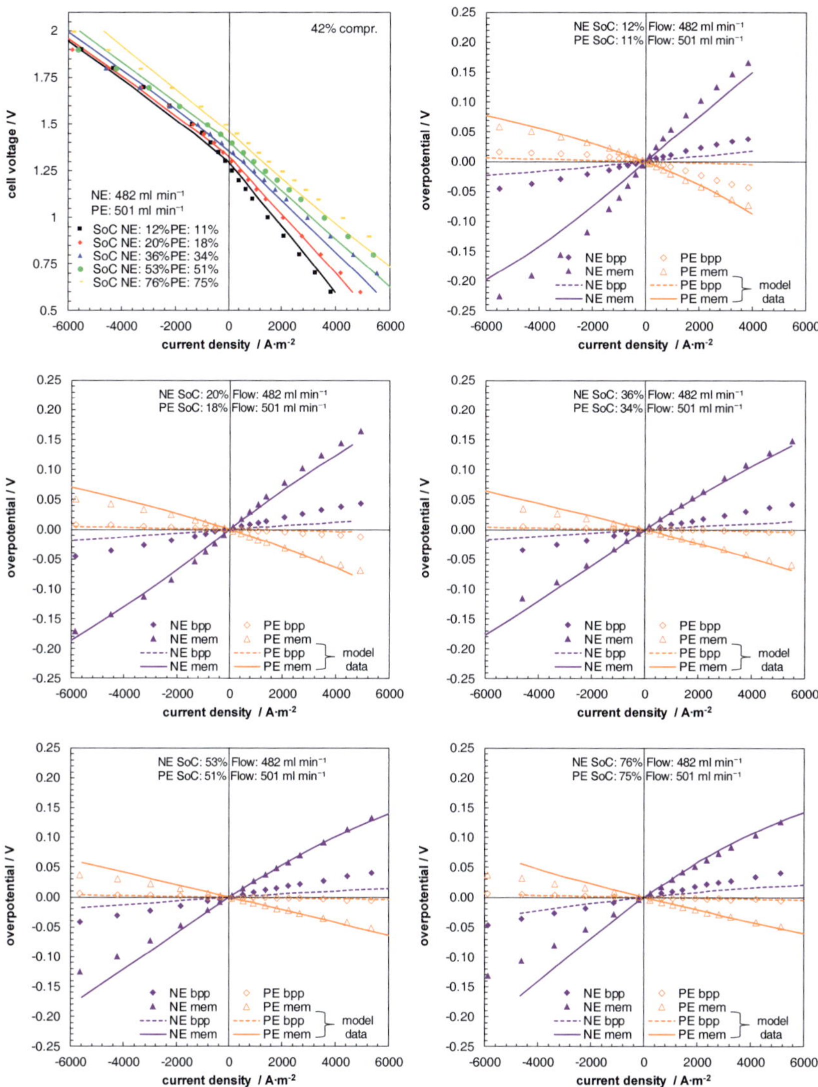

Figure A-12: Top left) Polarization curves of a cell setup with 42% carbon felt compression rate (GFD4.6E carbon felts, Nafion 117 membrane) for several states of charge. Others) Overpotentials measured at bipolar plate and membrane interface calculated by the difference between liquid and solid phase potential probe. Corresponding flow rate and SoC are shown in the figures for and PE, respectively. Experimental data represented as dots, whereas data from the model is indicated by lines.

A.2 gPROMS code

In order to be able to rebuild the same model in gPROMS®, the used codes for the electrode and the full cell are given in this section.

Model of the electrode

```
1   PARAMETER
2
3       # set of reactions
4       EcReactions as ORDERED_SET
5       # set of components
6       components as ORDERED_SET
7
8       # reaction coefficients and orders
9           # for mass balance
10      nue as array(EcReactions,components) of real default 0
11          # for anodic branch in Butler-Volmer-equation
12      order_an as array(EcReactions, components) of real default 0
13          # for cathodic branch in Butler-Volmer-equation
14      order_cat as array(EcReactions, components) of real default 0
15          # for calculation of exch. current dens.
16      order_io_plus, order_io_minus as array(EcReactions) of real default 0
17          # allows deactivation of single reactions if necessary
18      bool_ec as array(EcReactions) of integer default 1
19
20
21      # general parameters
22      D as array(components) of real
23      z as ARRAY(components) of integer
24      Rgas as real
25      T as real
26      F as real
27      cref as real
28
29      # hydraulic and felt diameter
30      dh_felt, d_fiber as real
31
32      # felt's properties
33      spec_area as real
34      porosity as real
35      permeability as real
36
37      # dimensions of the felt
38      height, width as real
39      thickness as real
40
41      DISTRIBUTION_DOMAIN
42      y_dim as  [0 : height]
43      x_dim as  [0 : thickness]
44
```

```
44
45
46   VARIABLE
47   PhiS as distribution( x_dim, y_dim) of potential
48   PhiL as distribution( x_dim, y_dim) of potential
49
50   # overpotential & redox_potential at surface and in the bulk
51   eta as distribution( EcReactions, x_dim, y_dim) of potential
52   Redoxpot, Redoxpot_surf as distribution( EcReactions, x_dim, y_dim) of potential
53
54   # current densities in solid and liquid phase and at the current collector
55   jS as distribution( x_dim, y_dim) of current_density
56   jL as distribution( x_dim, y_dim) of current_density
57   j as distribution (y_dim) of current_density
58
59   # transfer and exchange current density for a specific reactions
60   i,i0 as distribution( EcReactions, x_dim, y_dim) of current_density
61
62   # effective faradaic current at electrode (in case of multiple reactions)
63   ieff as distribution( x_dim, y_dim) of current_density
64
65   # sources and sinks due to electrochemical reactions
66   sourceEC[, sourceCH ]as DISTRIBUTION(components, x_dim, y_dim) of flux
67
68   # electrolyte concentrations in bulk, at surface & at the inlet
69   cbulk as distribution( components,  x_dim, y_dim) of concentration
70   csurf as distribution( components,  x_dim, y_dim) of concentration
71   c_in as array( components) of concentration
72
73   # inlet flow rate and resulting flux
74   Vin, qy as flux
75   # inlet pressure (not really used in the model)
76   inlet_pressure as pressure
77
78   # Reynolds, Sherwood and Schmidtnumber
79   Re, Sh,Sc as array (components) of value
80
81   # thickness of diffusion boundary layer
82   delta as array(components) of length
83   # Parameters for Sherwoodcorrelation
84   ShO, ShA, ShB, ShC as value
85
86   # electrochemical kinetic parameters
87   k0 as array(EcReactions) of value   #Reaktionsgeschwindigkeiteskonstanten
88   alpha_an as array(EcReactions) of value  # Symmetriefaktor anodisch
89   alpha_cat as array(EcReactions) of value# Symmetriefaktor kathodisch
90
```

```
 90
 91     # reference potentials
 92     Eref as array(EcReactions) of value
 93     # effective resistivities in solid and liquid phase
 94     electric_resistivity_S, electric_resistivity_L as value
 95     # electrolyte viscosity
 96     electrolyte_viscosity as value
 97
 98  SELECTOR
 99     mass_transport as (no_limitation, no_inner_limitation, real_transport) DEFAULT real_transport
100
101  BOUNDARY
102         for s in components do
103             cbulk(s , 0:thickness, 0) = c_in(s);
104         end
105
106  EQUATION
107 # Model equations
108             # electrolyte flux
109             qy = Vin/(width *thickness);
110             inlet_pressure = height * electrolyte viscosity / permeability * qy;
111
112             # Reynoldsnumber, Schmidt & Sherwoodnumber and resulting diffusion boundary layer
113             for s in components do
114             Re(s) = d_fiber /porosity * qy * 1350 / electrolyte_viscosity;
115             Sc(s) = electrolyte_viscosity/1350 /D(s);
116             Sh(s) = ShO+ShA * Re(s)^ShB * Sc(s)^ShC; # Sherwood
117             delta(s) = d_fiber/Sh(s);
118             end
119
120 ############# electrochemical reactions #############
121                 for y := 0 to height do
122                     for x := 0 to thickness do
123                         for r in EcReactions do
124                             # Redoxpotentials and overpotentials
125                             Redoxpot(r,x,y) = Eref(r)+
126                                     Rgas*T/F*(log(0.0 + product(cbulk(,x,y)^order_cat(r,)))
127                                     -log(0.0 + product(cbulk(,x,y)^order_an(r,))));
128                             Redoxpot_surf(r,x,y) = Eref(r)+
129                                     Rgas*T/F*(log(0.0 + product(csurf(,x,y)^order_cat(r,)))
130                                     -log(0.0+ product(csurf(,x,y)^order_an(r,))));
131                             eta(r,x,y) = PhiS(x,y)-PhiL(x,y)-Redoxpot_surf(r,x,y);
132                             # exchange current density
133                             i0(r,x,y) = F*k0(r)*cref *
134                                     product((csurf(,x,y)/cref)^order_cat(r,))^order_io_plus(r)  *
135                                     product((csurf(,x,y)/cref)^order_an (r,))^order_io_minus(r);
136                             # transfer current density
137                             i(r,x,y) = i0(r,x,y)*(exp(alpha_an(r)*eta(r,x,y)*F/(Rgas*T))
138                                     - exp(alpha_cat(r)*(-eta(r,x,y))*F/(Rgas*T)));
139                         end
```

```
140
141                       # sinks and sources
142                       for s in components do
143                           sourceEC(s,x,y) = sigma(nue(,s)*i(,x,y)) * spec_area/F;
144                       end
145                       # effective transfer current
146                       ieff(x,y) = Sigma(i(,x,y));
147                   end
148                   # potential gradients
149                   for x := 0|+ to thickness do #+
150                       partial(PhiS(x,y),x_dim) = -electric_resistivity_S*jS(x,y);
151                   end
152                   for x := 0 to thickness do #+
153                       partial(PhiL(x,y),x_dim) = -electric_resistivity_L*jL(x,y);
154                   end
155                   # gradients in current density in solid phase
156                   for x := 0|+ to thickness|- do #+
157                       partial(jS(x,y),x_dim) = -spec_area*ieff(x,y);
158                   end
159                   # gradients in current density in liquid phase
160                   for x:= 0|+ to thickness do #+
161                       jS(x,y) = j(y)-jL(x,y);
162                   end
163               end
164
165 # selectors for initialisation procedure
166  CASE mass_transport OF
167     WHEN no_limitation:
168               # assume all concentrations equal to inlet
169               for y := 0|+ to height do
170                   for x:= 0|+ to thickness|- do
171                       for s in components do
172                           cbulk(s,x,y) = c_in(s);
173                       end
174                   end
175               end
176               for y := 0 to height do
177                   for x:= 0 to thickness do
178                       for s in components do
179                           csurf(s,x,y) = cbulk(s,x,y);
180                       end
181                   end
182               end
183     WHEN no_inner_limitation:
```

```
183     WHEN no_inner_limitation:
184             # no mass transport limitation from bulk to surface
185         for y := 0|+ to height do
186             for x:= 0|+ to thickness|- do
187                 for s in components do
188                     porosity*$cbulk(s,x,y)*0 = sourceEC(s,x,y)
189                     - qy*partial(cbulk(s,x,y),y_dim);
190                 end
191             end
192         end
193         for y := 0 to height do
194             for x:= 0 to thickness do
195                 for s in components do
196                     csurf(s,x,y) = cbulk(s,x,y);
197                 end
198             end
199         end
200     WHEN real_transport:
201                 # all effects included
202         for y := 0|+ to height do
203             for x:= 0|+ to thickness|- do
204                 for s in components do
205                     porosity*$cbulk(s,x,y)*0 = sourceEC(s,x,y)-qy*partial(cbulk(s,x,y),y_dim);
206                 end
207             end
208         end
209         for y := 0 to height do
210             for x:= 0 to thickness do
211                 for s in components do
212                     csurf(s,x,y) = cbulk(s,x,y)+sigma(nue(,s)*i(,x,y))*(delta(s))/F/D(s);
213                 end
214             end
215         end
216 END
217
218 INITIALISATION_PROCEDURE IP_simpled_massbalance
219 START
220   mass_transport := no_limitation;
221 END
222
223 NEXT
224   MOVE_TO
225     mass_transport := no_inner_limitation;
226   END
227 END
228
229 NEXT
230   MOVE_TO
231     mass_transport := real_transport;
232   END
```

Model of the full cell

```
 1 | PARAMETER
 2     # General cell parameters
 3     height as real
 4     width as real
 5
 6     # thickness of bipolar plate
 7     dBpp as real
 8     # variable for the position of the probes within the cell
 9     Messhoehe as real
10
11 DISTRIBUTION_DOMAIN
12     y_dim as [0 : height]
13
14 UNIT
15     # two sub models for the NE and PE electrode
16     PE_Felt as Felt_2D_01
17     NE_Felt as Felt_2D_01
18
19 VARIABLE
20     # local current density over height of cell
21     j as DISTRIBUTION(y_dim) of current_density
22     # effective current density of the cell
23     jeff as current_density
24
25     # voltage drops across the bipolar plates and the membrane
26     Delta_PhiBppNE, Delta_PhiBppPE, Delta_PhiMem as Distribution(y_dim) of potential
27     # cell voltage
28     UCell as potential
29
30     # Parameters for mass transport
31     ShO, ShA, ShB, ShC as value
32
33     ##Potential probe signals measured in the cell
34     Phi_S_B_NE as potential
35     Phi_S_M_NE as potential
36     Phi_L_B_NE as potential
37     Phi_L_M_NE as potential
38     Phi_S_B_PE as potential
39     Phi_S_M_PE as potential
40     Phi_L_B_PE as potential
41     Phi_L_M_PE as potential
42
43     # electrical resistance of the bipolar plate and the felt
44     resBpp, resfelt as value
45     # resitances of the membrane
46     resMem_charge, resMeM_discharge as value
47
```

```
47
48     # inlet concentrations of the electrolyte
49     SoC_In_PE, SoC_In_NE as value
50     cV_PE, cV_NE as concentration
51
52  BOUNDARY
53 # Boundary condition equations
54
55         #effective current density is the integral of the distributed current density
56         jeff = integral( y := 0:height ; j(y) )/ height;
57         # define current densities in the NE submodel
58         NE_Felt.jL(0, 0:height) = 0;
59         NE_Felt.jS(0, 0:height) = j(0:height);
60         NE_Felt.jS(NE_felt.thickness, 0:height) = 0;
61         # definition of NE electrode boundary potential
62         NE_Felt.PhiS(0, 0:height) = {PhiS0 = 0}0 + Delta_PhiBppNE(0:height);
63         # no diffusion across at the NE electrode boundaries
64         Partial (NE_Felt.cbulk( , 0, 0|+:height), NE_Felt.x_dim) = 0;
65         Partial (NE_Felt.cbulk( , NE_felt.thickness, 0|+:height), NE_Felt.x_dim) = 0;
66
67         # liquid phase potential link between NE and PE
68         PE_Felt.PhiL(0, 0:height) = NE_Felt.PhiL(NE_Felt.thickness, 0:height) + Delta_PhiMem(0:height);
69         # define current densities in the PE submodel
70         PE_Felt.jS(0, 0:height) = 0;
71         PE_Felt.jS(PE_Felt.thickness, 0:height) = j(0:height);
72         PE_Felt.jL(0, 0:height) = j(0:height);
73         # no diffusion across at the PE electrode boundaries
74         Partial (PE_Felt.cbulk( , 0, 0|+:height), PE_Felt.x_dim) = 0;
75         Partial (PE_Felt.cbulk( , PE_felt.thickness, 0|+:height), PE_Felt.x_dim) = 0;
76
77         # calculation of the cell voltage
78         UCell = PE_Felt.PhiS(PE_Felt.thickness, 0:height) + Delta_PhiBppPE(0:height);
79
80         # calculate inlet concentrations for the sub models NE&PE
81         NE_Felt.c_in('V2') = cV_NE * SoC_In_NE;
82         NE_Felt.c_in('V3') = cV_NE * (1-SoC_In_NE);
83
84         PE_Felt.c_in('V5') = cV_PE * SoC_In_PE;
85         PE_Felt.c_in('V4') = cV_PE * (1-SoC_In_PE);
86
87         # link current densities to the sub models
88         PE_Felt.j(0:height) = j(0:height);
89         NE_Felt.j(0:height) = j(0:height);
90
```

```
90
91  EQUATION
92  # Model equations
93      FOR y:= 0 to height DO
94          # calculation of voltage drops across bipolar plates
95          Delta_PhiBppNE(y) = -j(y) * resBpp[_NE]{*dBpp};
96          Delta_PhiBppPE(y) = -j(y) * resBpp[_PE]{*dBpp};
97          # calculation of votlage drop across membrane
98          Delta_PhiMem(y) = -j(y)* (resMem_charge + (sgn(j(y))+1)/2*(resMeM_discharge-resMem_charge));
99
100     END
101
102
103         within NE_Felt do
104             # define resistivity in submodel
105             electric_resistivity_S = resfelt;
106             # calculate potential probe signals in the NE
107             Phi_S_B_NE = 0 - Delta_PhiBppNE(Messhoehe);
108             Phi_S_M_NE = 0 - PhiS(thickness, Messhoehe);
109             Phi_L_B_NE = 0 - PhiL(        0, Messhoehe) - Redoxpot('V23',           0, Messhoehe);
110             Phi_L_M_NE = 0 - PhiL(thickness, Messhoehe) - Redoxpot('V23', thickness, Messhoehe);
111         end
112
113         within PE_Felt do
114             # define resistivity in submodel
115             electric_resistivity_S = resfelt;  # Werte von FullCell auf Felt übertragen
116             # calculate potential probe signals in the PE
117             Phi_S_B_PE = Delta_PhiBppPE(Messhoehe);
118             Phi_S_M_PE = UCell - PhiS(0, Messhoehe);
119             Phi_L_B_PE = UCell - PhiL(thickness, Messhoehe) - Redoxpot('V45', thickness, Messhoehe);
120             Phi_L_M_PE = UCell - PhiL(        0, Messhoehe) - Redoxpot('V45',           0, Messhoehe);
121         end
122
123         #Defining mass transport parameters in the sub models
124         NE_Felt.ShO = ShO;
125         PE_Felt.ShO = ShO;
126
127         NE_Felt.ShA = ShA;
128         PE_Felt.ShA = ShA;
129
130         NE_Felt.ShB = ShB;
131         PE_Felt.ShB = ShB;
132
133         NE_Felt.ShC = ShC;
134         PE_Felt.ShC = ShC;
135
136         # calculation of membrane resistances
137         resMem_charge = (0.372 + (SoC_In_NE + SoC_In_PE)/2*0.088)/10000;
138         resMeM_discharge = (0.908 - (SoC_In_NE + SoC_In_PE)/2*0.354)/10000;
```

```
135
136           # calculation of membrane resistances
137           resMem_charge = (0.372 + (SoC_In_NE + SoC_In_PE)/2*0.088)/10000;
138           resMeM_discharge = (0.908 - (SoC_In_NE + SoC_In_PE)/2*0.354)/10000;
139
140
141           # calculation pf electrolyte resistivities
142           NE_Felt.electric_resistivity_L = 1/((12.0*SoC_In_NE + 22.8)*NE_Felt.porosity^1.5);
143           PE_Felt.electric_resistivity_L = 1/((13.3*SoC_In_PE + 30.9)*PE_Felt.porosity^1.5);
144
145           # calculation of electrolyte viscosities
146           NE_Felt.electrolyte_viscosity = (-1.85 * SoC_In_NE + 6.01)*1e-3;
147           PE_Felt.electrolyte_viscosity = (-0.76 * SoC_In_PE + 4.53)*1e-3;
148
149 # definition of the initialisation procedure
150 INITIALISATION_PROCEDURE start_fullcell
151     USE
152         [Felt_2D_01] : IP_simpled_massbalance;
153
154     END
155
```